ENCYCLOPAEDIA OF
NANO ENGINEERING

ENCYCLOPAEDIA OF
NANO ENGINEERING

Thomas Watts

ANMOL PUBLICATIONS PVT. LTD.
NEW DELHI - 110 002 (INDIA)

ANMOL PUBLICATIONS PVT. LTD.
H.O.: 4374/4B, Ansari Road, Darya Ganj,
New Delhi-110 002 (India)
Ph.: 23278000, 23261597
B.O.: No. 1015, Ist Main Road, BSK IIIrd Stage
IIIrd Phase, IIIrd Block,
Bangalore - 560 085 (India)
Visit us at: www.anmolpublications.com

Encyclopaedia of Nano Engineering

First Published, 2006
ISBN 81-261-2982-4

PRINTED IN INDIA

Printed at Mehra Offset Press, Delhi.

Contents

Preface

Understanding the relationship among nano-science, nanotechnology and nano-engineering is the main goal of this book. Nano Engineering primarily involves study of the process of self-assembly at a nanoscale. Opportunities and challenges presented by contemporary science of nano- microscopy, nano-engineering & nanomanipulation are presented. An overview of major applications and importance of nano-engineering is given in detail in this book. The industrial and other applications of nano-electromechanical systems have been described briefly. Subjects that seem at present to have the most direct and constructive involvement with nano-engineering are biomedical engineering (through genetics and microbiology); electrical engineering (through MEMS systems); and software engineering. In general, nanotechnology involves the manipulation of matter at nanometer length (one-billionth of a meter) scales to produce new materials, structures and devices. It may be defined as a technology, which involves research and technology development involving structures with at least one dimension in approximately the 1-100 nanometer range, frequently with atomic/molecular precision. It also involves processes for creating and using structures, devices and systems that have novel properties and functions because of their nanometer scale dimensions. The discovery in recent years of the mechanisms for accessing individual building blocks of matter, and with it the increasing understanding of how these building blocks organize themselves, has sparked off the industrial conquest of the nanosphere throughout the world.

This book entitled "Encyclopaedia of Nano Engineering" tends to cover the aforementioned subject areas in their broadest possible terms. All useful supplementary research and reference tools have been included for readers' further investigations in the subject area of Nano Engineering. This book mainly focuses on nano engineering in terms of its fundamentals and applications, nanoscience instruments, nano-electromagnetic systems, nanomechanics, nanomachines and self-Assembly. This book on Nano Engineering seeks to explain this complex field of study.

—Editor

1

Introduction to Nanotechnology & Nanoengineering

Nanotechnology: Consequences

The United States may have become famous for its obsession with the elections and the next quarter's profits, and the future be damned. Nonetheless, we are writing for normal human beings who feel that the future matters–ten, twenty, perhaps even thirty years from now—for people who care enough to try to shift the odds for the better. Making wise choices with an eye to the future requires a realistic picture of what the future can hold. What if most pictures of the future today are based on the wrong assumptions?

Here are a few of today's common assumptions, some so familiar that they are seldom stated:

- Industrial development is the only alternative to poverty.
- Many people must work in factories.
- Greater wealth means greater resource consumption.
- Logging, mining, and fossil fuel burning must continue.
- Manufacturing means polluting.
- Third World development would doom the environment.

These all depend on a more basic assumption:

- *Industry as we know it cannot be replaced.*

Some further common assumptions:

- Today's economic trends will define tomorrow's problems.
- Spaceflight will never be affordable for most people.

- Forests will never grow beyond Earth.
- More advanced medicine will always be more expensive.
- Even highly advanced medicine won't be able to keep people healthy.
- Solar energy will never become really inexpensive.
- Toxic wastes will never be gathered and eliminated.
- Developed land will never be returned to wilderness.
- There will never be weapons worse than nuclear missiles.
- Pollution and resource depletion will eventually bring war or collapse.

These, too, depend on a more basic assumption:

- *Technology as we know it will never be replaced.*

These commonplace assumptions paint a future full of terrible dilemmas, and the notion that a technological change will let us escape from them smacks of the idea that some technological fix can save the industrial system. The prospect, though, is quite different: The industrial system won't be fixed, it will be junked and recycled. The prospect isn't more industrial wealth ripped from the flesh of the Earth, but green wealth unfolding from processes as clean as a growing tree. Today, our industrial technologies force us to choose better quality or lower cost or greater safety or a cleaner environment. Molecular manufacturing, however, can be used to improve quality *and* lower costs *and* increase safety *and* clean the environment. The coming revolutions in technology will transcend many of the old, familiar dilemmas. And yes, they will bring fresh, equally terrible dilemmas.

Molecular nanotechnology will bring thorough and inexpensive control of the structure of matter. We need to understand molecular nanotechnology in order to understand the future capabilities of the human race. This will help us see the challenges ahead, and help us plan how best to conserve values, traditions, and ecosystems through effective policies and institutions. Likewise, it can help us see what today's events mean, including business opportunities and possibilities for action. We need a vision of where technology is leading because technology is a part of what human beings are, and will affect what we and our societies can become.

The consequences of the coming revolutions will depend on human actions. As always, new abilities will create new possibilities both for good and for ill. We will discuss both, focusing on how political and economic pressures can best be harnessed to achieve good ends. Our answers will not be satisfactory, but they are at least a beginning.

Molecular Nanotechnology as Exploratory Engineering

Studies of molecular nanotechnology are today in the exploratory engineering phase, and just beginning to move into engineering development. The basic idea of exploratory engineering is simple: combine engineering principles with known scientific facts to form a picture of future technological possibilities. Exploratory engineering looks at future possibilities to help guide our attention in the present. Science–especially molecular science–has moved fast in recent decades. There is no need to wait for more scientific breakthroughs in order to make engineering breakthroughs in nanotechnology.

The above illustration shows how exploratory engineering relates to more familiar kinds of engineering. Each works within the limits of the possible, which are set by the known and unknown laws of nature. The most familiar kind is the engineering taught in schools: this "textbook engineering" covers technologies that can be both understood (so they can be taught) and manufactured (so they can be used). Bridge-building and gearbox design fall in this category. Other technologies, however, can be manufactured but aren't understood—any engineer can give examples of things that work when similar things don't, and for no obvious reason. But as long as they do work, and work consistently, they can be used with confidence. This is the world of "cut-and-try engineering," so important to modern industry. Bearing lubrication, adhesives, and many manufacturing technologies advance by cut-and-try methods. Exploratory engineering covers technologies that can be understood but not manufactured–yet. Technologies in this category are also familiar to engineers, although normally they design such things only for fun. So much is known about mechanics, thermodynamics, electronics, and so forth that engineers can often calculate what something will do, just from a description of it. Yet there is no reason why everything that can be correctly described must be manufacturable—the constraints are different. Exploratory engineering is as simple as textbook engineering, but neither military planners nor corporate executives see much profit in it, so it hasn't received much attention. The concepts of molecular manufacturing and molecular nanotechnology are straightforward results of exploratory-engineering research applied to molecular systems. As we observed above, the basic ideas could have been worked out forty years ago, if anyone had bothered. Naturally enough, both scientists and engineers were preoccupied with more immediate concerns. But now, with the threshold of nanotechnology approaching, attention is beginning to focus on where the next steps lead. Nanotechnology seems to be where the world is headed if technology keeps advancing, and competition practically guarantees that advances will continue. It will open both a huge range of opportunities for benefit and a huge range of opportunities for misuse. We will paint scenarios to give a sense of the prospects and possibilities, but we don't offer predictions of what will happen. Actual human choices and blunders will depend on a range of factors and alternatives beyond what we can hope to anticipate.

Nanostructure Science and Technology

Nanostructure science and technology is a broad and interdisciplinary area of research and development activity that has been growing explosively worldwide in the past few years. It has the potential for revolutionizing the ways in which materials and products are created and the range and nature of functionalities that can be accessed. It is already having a significant commercial impact, which will assuredly increase in the future. A worldwide study of research and development status and trends in nanoparticles, nanostructured materials, and nanodevices (or more concisely, nanostructure science and technology) was carried out during the period 1996-98 by an eight-person panel under the auspices of the World Technology (WTEC) Division of Loyola College. Led by the National Science Foundation, a wide range of U.S. government agencies commissioned and funded this study: the Air Force Office of Scientific Research, the Office of Naval Research, the Department of Commerce (including the National Institute of Standards and Technology and the Technology Administration), the Department of Energy, the National Institutes of Health, and the National Aeronautics and Space Administration. Their support indicates the breadth of interest in and the far-reaching potential of this burgeoning new field. The purpose of the study was to assess the current status and future trends internationally in research and development in the broad and rapidly growing area of nanostructure science and technology. The goals were fourfold:

1. provide the worldwide science and engineering community with a broadly inclusive and critical view of this field
2. identify promising areas for future research and commercial development
3. help stimulate development of an interdisciplinary international community of nanostructure researchers
4. encourage and identify opportunities for international collaboration

This report is the principal volume in a three-part publication of the activities and findings of the WTEC panel; it is an overview of the panel's observations and conclusions regarding nanostructure science and technology worldwide. It includes reviews of panel workshops held in Germany and Sweden, as well as site reports of panel visits to university, government, and industry laboratories in Europe, Japan, and Taiwan. An earlier volume, published in January 1998, reported the proceedings of a WTEC workshop on R&D Status and Trends in Nanoparticles, Nanostructured Materials, and Nanodevices in the United States (Baltimore: Loyola College, International Technology Research Institute, NTIS #PB98- 117914). A third volume to be published by WTEC reports the proceedings of a workshop held in St. Petersburg, Russia on related work.

Findings

There are two overarching findings from this WTEC study:

First, it is abundantly clear that we are now able to nanostructure materials for novel performance. That is the essential theme of this field: novel performance through nanostructuring. It represents the beginning of a revolutionary new age in our ability to manipulate materials for the good of humanity. The synthesis and control of materials in nanometer dimensions can access new material properties and device characteristics in unprecedented ways, and work is rapidly expanding worldwide in exploiting the opportunities offered through nanostructuring. Each year sees an ever increasing number of researchers from a wide variety of disciplines enter the field, and each year sees an ever increasing breadth of novel ideas and exciting new opportunities explode on the international nanostructure scene.

Second, there is a very wide range of disciplines contributing to the developments in nanostructure science and technology worldwide. The rapidly increasing level of interdisciplinary activity in nanostructuring is exciting and growing in importance, and the intersections between the various disciplines are where much of the novel activity resides. The field of nanostructure science and technology has been growing very rapidly in the past few years, since the realization that creating new materials and devices from nanoscale building blocks could access new and improved properties and functionalities. While many aspects of the field existed well before nanostructure science and technology became a definable entity in the past decade, it has only become a coherent field of endeavor through the confluence of three important technological streams:

1. new and improved control of the size and manipulation of nanoscale building blocks
2. new and improved characterization (spatial resolution, chemical sensitivity, etc.) of materials at the nanoscale
3. new and improved understanding of the relationships between nanostructure and properties and how these can be engineered

As a result of these developments, a wide range of new opportunities for research and applications in the field of nanotechnology now present themselves. Some examples of present and potential applications with significant technological impact that were identified in the course of this study. Considerable resources are being expended around the world for research and development aimed at realizing these and a variety of other promising applications. Government funding alone approached half a billion dollars per year in FY 1997: $128 million in Western Europe; $120 million in Japan; $116 million in the United States; and $70 million altogether in other countries such as China, Canada, Australia, Korea, Taiwan, Singapore, and the countries of the former Soviet Union.

An overall comparison of the current levels of activity among the major regions assessed (Europe, Japan, and the United States) in the various areas of the WTEC study. These broad areas— synthesis and assembly, biological approaches and applications, dispersions and coatings, high surface area materials, nanodevices, and consolidated materials—constitute the field of nanostructure science and technology. These are the areas around which the study was crafted. In the synthesis and assembly area, the United States appears to be ahead, with Europe following and then Japan. In the area of biological approaches and applications, the United States and Europe appear to be rather on a par, with Japan following. In nanoscale dispersions and coatings, the United States and Europe are again at a similar level, with Japan following. For high surface area materials, the United States is clearly ahead of Europe, followed by Japan. On the other hand, in the nanodevices area, Japan seems to be leading quite strongly, with Europe and the United States following. And finally, in the area of consolidated materials, Japan is a clear leader, with the United States and Europe following. These comparisons are, of course, integrals over rather large areas of a huge field and therefore possess all of the inevitable faults of such an integration. At best, they represent only a snapshot of the present, and the picture is admittedly incomplete.

More detailed findings in each of these major areas are included in the individual chapters of this report, along with additional general findings and observations compares the scope and funding levels for the relevant nanostructure science and technology R&D programs around the world. The appendices give details on the site visits and workshops of the panel: B contains the Europe site reports, C contains notes on workshops held in Germany and Sweden, D contains the Japan site reports, and E contains the Taiwan site reports.

Challenges

We are now at the threshold of a revolution in the ways in which materials and products are created. How this revolution will develop, how great will be the opportunities that nanostructuring can yield, and how rapidly we progress, will depend upon the ways in which a number of challenges are met.

Among the challenges facing us are those concerned with making the necessary advances in enabling technologies in order for rapid progress to continue in this field. We must increase characterization capabilities in visualization and chemical analysis at ever finer size scales. We must be able to manipulate matter at ever finer size scales, and we must eventually use computational approaches in directing this, if we are really going to take full advantage of the available opportunities. Experiment simply cannot do it alone. Theory and modeling are essential. Fortunately, this is an area in which the sizes of the building blocks and their assemblies are small enough that one can, with the ever increasing capabilities in computational sciences, now start doing very serious controlled modeling experiments to guide us in the nanostructuring of matter. Hence, multiscale modeling of nanostructuring and the resulting materials properties across the hierarchy of length scales from atomic, to mesoscopic, to macroscopic is an absolute necessity as we go down the road in the next decades to realizing the tremendous potential of nanostructure science and technology.

Furthermore, we need to understand the critical roles that surfaces and interfaces play in nanostructured materials. Nanoparticles have very high specific surface areas, and thus in their assembled forms there are large areas of interfaces. We need to know in detail not only the structures of these interfaces, but also their local chemistries and the effects of segregation and interaction between the nanoscale building blocks and their surroundings. We need to learn more about the control of nanostructure size and size distribution, composition, and assembly. For some applications, there are very stringent conditions on these parameters; in other applications less so. We must therefore understand the relationships between this stringency and the desired material or device properties.

We also need to be concerned with the thermal, chemical, and structural stability of nanostructured materials and the devices made therefrom, in the face of both the temperature and changing chemistries of the environments in which these nanostructures are asked to function. A nanostructure that is only a nanostructure at the beginning of a process is not of much use to anybody, unless the process is over in a very short time or the process itself is the actual nanostructure advantage. So for many applications, stability is an important consideration, and we must investigate whether natural stability is sufficient or whether we must additionally stabilize against changes that we cannot afford.

To effectively commercialize and utilize the nanostructuring of matter we also need enhancements in statistically driven process control. Achieving reproducibility and scalability of nanoparticle synthesis and consolidation processes in nanostructuring are paramount if successful scale-up is to be effected and if what we do in the laboratory is to contribute to the society that pays for this research. Given a commercial need, the viability of nanostructure production and utilization is wrapped up in the costs of precursors or raw materials, processing costs, and also the costs of dealing with effluent. It is the total integrated cost, in terms of raw materials, synthesis of the building blocks, manufacture of parts from those building blocks, and effluent clean-up costs, that is important and that will ultimately determine commercial viability.

Finally, in order for the field of nanostructure science and technology to truly reach fruition, it is an absolute necessity to create a new breed of researchers who can work across traditional disciplines and think "outside the box." Educating this new breed of researchers, who will either work across disciplines or know how to work with others in the interfaces between disciplines, is vital to the future of nanostructure science and technology. People must start thinking in unconventional ways if we are to take full advantage of the opportunities in this new and revolutionary field.

Nanotechnology and Semiconductors

Semiconductors are:

(i) Solid crystalline materials whose electrical conductivities are intermediate between that of conductors and insulators, and which are usually strongly temperature-dependent.

(ii) A semiconductor is a material that is an insulator at very low temperature, but which has a sizable electrical conductivity at room temperature. The distinction between a semiconductor and an insulator is not very well-defined, but roughly, a semiconductor is an insulator with a band gap small enough that its conduction band is appreciably thermally populated at room temperature.

Before looking at how some of the semiconductor devices work it is necessary to have a basic understanding of what conductors, insulators and naturally what semiconductors are.

In terms of electricity there are two main classes of material: namely: conductors and non-conductors (or insulators). From their names it can be gathered that conductors will conduct electricity freely, whereas non-conductors act as insulators preventing the flow of an electric current. An electric current is made up of the flow of electrons. This means that for a current to flow, the electrons must be able to move freely within the material. In some materials electrons are moving freely around the lattice, not particularly attached to a given electron. At any instance electrons are moving freely but randomly. By placing a potential difference across the conductor the electrons can be made to drift in one direction and this constitutes an electric current. Metals are all conductors of electricity, and a number of other substances also conduct it to varying degrees. Other substances do not have electrons moving freely around the lattice. Electrons are firmly held within their molecules and cannot escape easily. Accordingly when a potential is placed across the substance very few electrons will move and very little or no current will flow. These substances are called non-conductors or insulators. They include most plastics, ceramics and many naturally occurring substances like wood.

In the parlance of solid-state physics, semiconductors (and insulators) are defined as solids in which at 0°K (and without excitations) the uppermost band of occupied electron energy states is completely full. It is well-known from solid-state physics that electrical conduction in solids occurs only via electrons in partially-filled bands, so conduction in pure semiconductors occurs only when electrons have been excited—thermally, optically, etc.—into higher unfilled bands. At room temperature, a proportion (generally very small, but not negligible) of electrons in a semiconductor have been thermally excited from the "valence band," the band filled at 0°K, to the "conduction band," the next higher band. The ease with which electrons can be excited from the valence band to the conduction band depends on the energy gap between the bands, and it is the size of this energy bandgap that serves as an arbitrary dividing line between semiconductors and insulators. Semiconductors generally have bandgaps of approximately 1 electron-volt, while insulators have bandgaps several times greater. When electrons are excited from the valence band to the conduction band in a semiconductor, both bands contribute to conduction, because electrical conduction can occur in any partially-filled energy band. The current-carrying electrons in the conduction band are known as "free electrons," though often they are simply called "electrons" if context allows this usage to be clear. The free energy-states in the valence band are known as "holes." It can be shown that holes behave very much like positively-charged counterparts of electrons, and they are usually treated as if they are real charged particles.

One of the main reasons that semiconductors are useful in electronics is that their electronic properties can be greatly altered in a controllable way by adding small amounts of impurities. These impurities, called dopants, add extra electrons or holes. A semiconductor with extra electrons is called an n-type semiconductor, while a semiconductor with extra holes is called a p-type semiconductor. The most common n-type dopants for silicon are phosphorus and arsenic. Notice that these two elements are in Group V of the periodic table, and silicon is in Group IV. When silicon is doped with arsenic or phosphorus atoms, these dopant atoms replace silicon atoms in the semiconductor crystal, but since they have one more outer-shell electron than silicon they tend to contribute this electron to the conduction band. By far the most common p-type dopants for silicon is the Group III element boron, which lacks an outer-shell electron compared with silicon and thus tends to contribute a hole to the valence band.

Heavily doping a semiconductor can increase its conductivity by a factor greater than a billion. In modern integrated circuits, for instance, heavily-doped polycrystalline silicon is often used as a replacement for metals.

An intrinsic semiconductor is one which is pure enough that impurities do not appreciably affect its electrical behavior. In this case, all carriers are created by thermally or optically exciting electrons from the full valence band into the empty conduction band. Thus, equal numbers of electrons and holes are present in an intrinsic semiconductor. Electrons and holes flow in opposite directions in an electric field, though they contribute to current in the same direction since they are oppositely charged. Hole current and electron current are not necessarily equal in an intrinsic semiconductor, however, because electrons and holes have different effective masses (crystalline analogues to free inertial masses). The concentration of carriers is strongly dependent on the temperature. At low temperatures, the valence band is completely full, making the material an insulator (see electrical conduction for more information). Increasing the temperature leads to an increase in the number of carriers and a corresponding increase in conductivity. This principle is used in thermistors. This behavior contrasts sharply with that of most metals, which tend to become less conductive at higher temperatures due to increased phonon scattering. An extrinsic semiconductor is one that has been doped with impurities to modify the number and type of free charge carriers.

The purpose of n-type doping is to produce an abundance of carrier electrons in the material. To help understand how n-type doping is accomplished, consider the case of silicon (Si). Si atoms have four valence electrons, each of which is covalently bonded with one of four adjacent Si atoms. If an atom with five valence electrons, such as those from group VA of the periodic table (eg. phosphorus (P), arsenic (As), or antimony (Sb)), is incorporated into the crystal lattice in place of a Si atom, then that atom will have four covalent bonds and one unbonded electron. This extra electron is only weakly bound to the atom and can easily be excited into the conduction band. At normal temperatures, virtually all such electrons are excited into the conduction band. Since excitation of these electrons does not result in the formation of a hole, the number of electrons in such a material far exceeds the number of holes. In this case the electrons are the majority carriers and the holes are the minority carriers. Because the five-electron atoms have an extra electron to "donate", they are called donor atoms.

The purpose of p-type doping is to create an abundance of holes. In the case of silicon a trivalent atom, such as boron, is substituted into the crystal lattice. The result is that an electron is missing from one of the four possible covalent bonds. Thus the atom can accept an electron from the valence band to complete the fourth bond, resulting in the formation of a hole. Such dopants are called acceptors. When a sufficiently large number of acceptors are added, the holes greatly outnumber the excited electrons. Thus, the holes are the majority carriers, while electrons are the minority carriers in p-type materials. Blue diamonds (Type IIb), which contain boron impurities, are an example of a naturally occurring p-type semiconductor.

A p-n junction may be created by doping adjacent regions of a semiconductor with p-type and n-type dopants. If a positive bias voltage is placed on the p-type side, the dominant positive carriers (holes) are pushed toward the junction. At the same time, the dominant negative carriers (electrons) in the n-type material are attracted toward the junction. Since there is an abundance of carriers at the junction, current can flow through the junction from a power supply, such as a battery. However, if the bias is reversed, the holes and electrons are pulled away from the junction, leaving a region of relatively non-conducting silicon which inhibits current flow. The p-n junction is the basis of an electronic device called a diode, which allows electric current to flow in only one direction. Similarly, a third region can be doped n-type or p-type to form a three-terminal device, such as the bipolar junction transistor (which can be either p-n-p or n-p-n).

Semiconductors with predictable, reliable electronic properties are difficult to mass-produce because of the required chemical purity, and the perfection of the crystal structure, which are needed to make devices. Because the presence of impurities in very small proportions can have such big effects on the properties of the material, the level of chemical purity needed is extremely high. Techniques for achieving such high purity include zone refining, in which part of a solid crystal is melted. Impurities tend to concentrate in the melted region, leaving the solid material more pure. A high degree of crystalline perfection is also required, since faults in crystal structure such as dislocations, twins, and stacking faults, create energy levels in the band gap, interfering with the electronic properties of the material. Faults like these are a major cause of defective devices in production processes. The larger the crystal, the harder it is to achieve the necessary purity and perfection; current mass production processes use six-inch diameter crystals which are grown as cylinders and sliced into wafers.

Semiconductor devices are electronic components that exploit the electronic properties of semiconductor materials, principally silicon, germanium and gallium arsenide. Semiconductor devices have replaced thermionic devices in most applications. They utilize electronic conduction in the solid state, as opposed to the vacuum state or gaseous state. Semiconductor devices are available as discrete units (such as those at Radio Shack) or can be integrated along with a large number—often millions—of similar devices onto a single chip, called an integrated circuit (IC).

If a semiconductor is pure and if it is unexcited by an input like an electric field, it allows very little current to pass through it, and it is practically an insulator. The main reason that semiconductors are so useful is that the conductivity of semiconductors can be manipulated by addition of impurities (doping), by introduction of an electric field, by exposure to light, or by other means. For example, CCDs, the primary unit of digital cameras, rely on the fact that semiconductor conductivity increases with exposure to light. Transistor operation, which will be discussed below, depends on the fact that semiconductor conductivity can be increased by the presence of an electric field. Current conduction in a semiconductor occurs via "free electrons" and "holes." Holes aren't real particles; in a sense that requires some knowledge of semiconductor physics to understand, a hole is the absence of an electron. Nevertheless, this absence, or hole, can be treated as a positively-charged counterpart to the negatively-charged electron. Indeed, the precise meaning of "free electrons" also requires a background in semiconductor physics to understand. For descriptive ease, "free electrons" are often simply denoted "electrons," but it should be understood that the majority of electrons in a solid, which aren't free, do not contribute to conductivity. If a semiconductor crystal is perfectly pure, with no impurities, and it is held at a temperature near absolute zero with no excitations (e.g. electric fields or light), it will contain no free electrons and no holes, and thus will be a perfect insulator. At room temperature, thermal excitations produce some free electrons and holes in pairs, but most semiconductors at room temperature are insulators for practical purposes. Doping a semiconductor, like silicon, with impurity atoms, like boron and phosphorus creates unequal numbers of free electrons and holes. High levels of doping can make a semiconductor a good conductor. When a doped semiconductor contains excess holes it is called "p-type," and when it contains excess free electrons it is known as "n-type." The semiconducting material in devices is almost always carefully doped for engineering purposes. In fact, junctions between n-type and p-type semiconductors, called p-n junctions, are the fundamental elements of many semiconductor devices, such as the p-n diode and the bipolar junction transistor. An electric field can also create a unequal number of free electrons and holes in a semiconductor. This is the basis for "field effect transistors" like the MOSFET. Exposure to light generally creates electron/ hole pairs in a semiconductor, which change its conductivity and allows the light to be sensed.

The most important semiconductor device in use today is the MOSFET, a type of transistor whose operation depends on a "gate," from which originates an electric field that controls the conductivity of a "channel." MOSFETs, as well as other transistor types, are used as the building blocks of logic gates. Their role in a microprocessor is somewhat analogous to that of neurons in the brain. In digital circuits like microprocessors, transistors act as on-off switches; in the MOSFET, for instance, the input gate voltage determines whether the switch is on or off. Transistors also serve very crucial operations in analog circuits. Transistors used for analog purposes do not act as on-off switches; rather, they respond to a continuous range of inputs with a continuous range of outputs. For example, transistors are the workhorses of amplifier circuits, which produce an amplified, but otherwise identical (ideally), version of the input. For years the bipolar junction transistor, or BJT, was the transistor of choice for analog circuits. However, the MOSFET has much more desirable properties for digital circuits, and since it is difficult to integrate BJTs and MOSFETs onto a single chip, MOSFETs are now commonly used for both analog and digital purposes.

While the transistor is the most important semiconductor device, there are literally dozens of other families of semiconductor devices. Some of the great many semiconductor devices in current usage are as follows:

Two-terminal devices:

- Semiconductor diode
- Zener diode
- Light-emitting diode
- PIN diode
- Schottky diode
- Avalanche diode
- Laser diode
- DIAC

Three-terminal devices:

- Bipolar transistor
- Field effect transistor

- Darlington transistor
- Thyristor
- Triac

Four-terminal devices:

- Hall effect sensor (magnetic field sensor)

The type designators of semiconductor devices are often manufacturer specific. Nevertheless, there have been attempts at creating standards for type codes, and a subset of devices follow those. For discrete devices, for example, there are three standards: JEDEC JESD370B in USA, Pro Electron in Europe and JIS in Japan.

Semiconductor device fabrication is a multiple-step sequence of predominantly chemical processing steps to create chips used in everyday electrical and electronic devices. Silicon is the most common semiconductor used today, although gallium arsenide, germanium, and many other materials are used in special applications.

To create the standard silicon-based chips that are used in most computers, raw silicon dioxide, which is the most common part of sand or glass, is heated in the presence of hydrogen to produce pure silicon and water which is extracted. The silicon is then melted and formed into ingots, that are sliced into wafers approximately 0.5mm thick and ranging from a few centimeters to tens of centimeters in diameter.

Once the wafers are prepared, a large number of process steps are necessary to produce the desired semiconductor integrated circuits. In general the steps can be grouped into four areas: Front End Processing, Back End Processing, Test and Packaging.

Front End Processing refers to the most crucial steps in the fabrication. In this stage the actual semiconductor devices or transistors are created. A typical front end process includes the following: preparation of the wafer surface, patterning and subsequent implantation of dopants to obtain the desired electrical properties, growth or deposition of a gate dielectric, and growth or deposition of insulating materials to isolate neighboring devices.

Once the various semiconductor devices have been created they must be interconnected to form the desired electrical circuits. This "Back End Processing" involves depositing various layers of metal and insulating material in the desired pattern. Typically the metal layers consist of aluminum or more recently copper. The insulating material was traditionally a form of SiO_2 or a silicate glass, but recently new low-K materials are being used. The various metal layers are interconnected by etching holes, called "vias" in the insulating material and depositing tungsten in them.

Once the Back End Processing has been completed, the semiconductor devices are subjected to a variety of electrical tests to determine if they function properly.

Finally, the wafer is cut into individual die, which are then packaged in ceramic or plastic packages with pins or other connectors to the outside world. Wafer Fabrication:

- Wet cleans
- Photolithography
- Ion implantation (in which dopants are embedded in the wafer creating regions of increased (or decreased) conductivity)
- Dry Etching
- Wet Etching
- Plasma ashing
- Thermal treatments
    - — Rapid thermal anneal
    - — Furnace anneals
    - — Oxidation
- Chemical vapor deposition (CVD)
- Physical vapor deposition (PVD)
- Molecular beam epitaxy (MBE)
- Electroplating
- Chemical mechanical polish (CMP)
- Wafer testing (where the electrical performance is verified)
- Wafer backgrinding (to reduce the thickness of the wafer so the resulting chip can be put into a thin device like a smartcard or PCMCIA card.)

Die preparation

- Wafer mounting
- Die cutting

IC packaging

- Die attaching

IC bonding

- Wire bonding
- Flip chip
- Tab bonding

IC encapsulation

- Electronic design automation
- GDSII
- OASIS

Nano Diode

A diode functions as the electronic version of a one-way valve. By restricting the direction of movement of charge carriers, it allows an electric current to flow in one direction, but blocks it in the opposite direction.

The first use for the diode was the demodulation of amplitude modulated (AM) radio broadcasts. The history of this discovery is treated in depth in the radio article. In summary, an AM signal consists of alternating positive and negative peaks of current, whose amplitude or 'envelope' is proportional to the original audio signal, but whose average value is zero. The diode rectifies the AM signal (i.e. it eliminates the negative peaks), leaving a signal whose average amplitude is the desired audio signal. The average value is extracted using a simple filter and fed into a transducer (originally a crystal earpiece, now more likely to be a loudspeaker), which generates sound. Diodes can be used to construct logic gates: logical and and logical or.

A diode is called a half wave rectifier when it is used to convert alternating current electricity into direct current, by removing the negative portion of the current. A special arrangement of four diodes that will transform an alternating current into a direct current, using both positive and negative excursions of a single phase alternating current, is known as a diode bridge, single-phase bridge rectifier, or simply a full wave rectifier.

With a split (center-tapped) alternating current supply it is possible to obtain full wave rectification with only two diodes. Often diodes come in pairs, as double diodes in the same housing. When it is desired to rectify three phase power, one could rectify each of the three phases with the arrangement of four diodes used in single phase, which would require a total of 12 diodes. However, due to redundancy, only six diodes are needed to make a three phase full wave rectifier. Most devices that generate alternating current (such devices are called alternators) generate three phase alternating current. Disassembled automobile alternator, showing the six diodes that comprise a full-wave three phase bridge rectifier. For example, an automobile alternator has six diodes inside it to function as a full wave rectifier for battery charge applications. Many of the small wind turbines use three double diodes bolted to the same heatsink.

Diodes are frequently used to conduct dangerously high voltages away from sensitive devices, most commonly by being reverse-biased (non-conducting) under normal circumstances, and becoming forward-biased (conducting) when the voltage rises above its normal value. For example, diodes are used in stepper motor and relay circuits to de-energize coils rapidly without the damaging voltage spikes that would otherwise occur. Many integrated circuits also incorporate diodes on the connection pins to prevent external voltages from damaging their sensitive transistors. Specialized diodes are used to protect from over-voltages at higher power.

The first diodes were vacuum tube devices (also known as thermionic valves), arrangements of electrodes surrounded by a vacuum within a glass envelope, similar in appearance to incandescent light bulbs. The arrangement of a filament and plate as a diode was invented in 1904 by John Ambrose Fleming, scientific adviser to the Marconi company, based on an observation by Thomas Edison. Like light bulbs, vacuum tube diodes have a filament through which current is passed, heating the filament. In its heated state it can now emit electrons into the vacuum. These electrons are electrostatically drawn to a positively charged outer metal plate called the anode, or just the "plate". Electrons do not flow from the plate back toward the filament, even if the charge on the plate is made negative, because the plate is not heated. Although vacuum tube diodes are still used for a few specialized applications, most modern diodes are based on semiconductor p-n junctions. In a p-n diode, conventional current can flow from the p-doped side (the anode) to the n-doped side (the cathode), but not in the opposite direction. When the diode is reverse-biased, the charge carriers are pulled away from the center of the device, creating a depletion region.

The gold causes 'minority carrier suppression.' This lowers the effective capacitance of the diode, allowing it to operate at signal frequencies. A typical example is the 1N914. Germanium and Schottky diodes are also fast like this, as are bipolar transistors 'degenerated' to act as diodes. Power supply diodes are made with the expectation of working at a maximum of 2.5 × 400 Hz (sometimes called 'French power' by Americans), and so are not useful above a kilohertz. Diodes that can be made to conduct backwards. This effect, called Zener Breakdown, occurs at a precisely defined voltage, allowing the diode to be used as a precision voltage reference. Some devices labelled as high-voltage Zener diodes are actually avalanche diodes. Two (equivalent) Zeners in series and in reverse order, in the same package, constitute a transient absorber (or *Transorb*, a registered trademark). They are named for Dr. Clarence Melvin Zener of Southern Illinois University, inventor of the device.

Diodes that conduct in the reverse direction when the reverse bias voltage exceeds the breakdown voltage. These are electrically very similar to Zener diodes, and are often mistakenly called Zener diodes, but break down by a different mechanism, the Avalanche Effect. This occurs when the reverse electric field across the p-n junction causes a wave of ionization, reminiscent of an avalanche, leading to a large current. Avalanche diodes are designed to break down at a well-defined reverse voltage without being destroyed. The difference between the avalanche diode (which has a reverse breakdown above about 6.2 V) and the Zener is that the channel length of the former exceeds the 'mean free path' of the electrons, so there are collisions between them on the way out. The only practical difference is that the two types have temperature coefficients of opposite polarities. Practical voltage reference circuits feature Zener and switching diodes connected in series and opposite directions to balance the temperature coefficient to near zero.

These are avalanche diodes designed specifically to protect other semiconductor devices from electrostatic discharges. Their p-n junctions have a much larger cross-sectional area than those of a normal diode, allowing them to conduct large currents to ground without sustaining damage.

As the electrons cross the junction they emit photons. In most diodes, these are reabsorbed, and are at frequencies that can not be seen (usually infrared). However, with the right materials and geometry, the light becomes visible. The forward potential of these diodes define their color. Thus different materials (extrinsic semiconductors) must be used. 1.2 V corresponds to red, 2.4 to violet. Now, even soft UV diodes are available. The first LED's were red and yellow, and higher-frequency diodes have been developed over time. Polishing the device with parallel faces, so as to form a resonant cavity, yields a 'laser diode.' All LEDs are monochromatic; 'white' LED's are actually combinations of three LED's of a different color, or a blue LED with a yellow scintillator coating. The lower the frequency of emission, the greater the efficiency. So to normalize output when using LED's of different colors, increase current in the higher frequency man eye is most sensitive in the blue-green.

These have wide, transparent junctions. Photons can push electrons over the junction, causing a current to flow. Photo diodes can be used as solar cells. And in photometry. If a photon doesn't have enough energy, it isn't going to turn the photodiode on very much. LED's can be used as low-efficiency photodiodes in signal applications. Sometimes a LED is paired with a photodiode or phototransistor in the same package. This device is called an "opto isolator." Unlike a transformer, this scheme allows for DC coupling. These are used to protect hospital patients from shock. Patients with IV's in their bodies are particularly susceptable, sometimes succumbing to 'carpet shock.' They are also used to isolate low-current control or signal circuitry from "dirty" power supply circuits or higher-current motor and machine circuits.

These have a very low forward voltage drop, usually 0.15 to 0.45 V, which makes them useful in battery-powered and low-voltage circuits. Also in mixer circuits for RF. These can provide very fast voltage transitions. These have a region of operation showing negative resistance caused by quantum tunneling, thus allowing amplification of signals and very simple bistable circuits.

These are similar to tunnel diodes in that they are made of materials such as GaAs or InP that exhibit a region of negative differential resistance. With appropriate biasing, dipole domains form and travel across the diode, allowing high frequency microwave oscillators to be built. There are other types of diodes, which all share the basic function of allowing electrical current to flow in only one direction, but with different methods of construction.

This works the same as the junction semiconductor diodes described above, but its construction is simpler. A block of n-type semiconductor is built, and a conducting sharp-point contact made with some group-3 metal is placed in contact with the semiconductor. Some metal migrates into the semiconductor to make a small region of p-type semiconductor near the contact. The long-popular 1N34 germanium version is still used in radio receivers as a detector and occasionally in specialized analog electronics.

This is the simplest kind of vacuum tube device (referred to as a valve in the UK). Electrons will move from a heated metal surface (cathode) treated with a mixture of barium and strontium oxides into a vacuum (thermionic emission). After leaving the cathode, they can be attracted to positively charged cool surface (anode). However, electrons are not easily released from a cold untreated surface when the voltage polarity is reversed and hence any flow is a very small current. For much of the 20th century they were used in analog signal applications, and as rectifiers in power supplies. Tube diodes were nearly obsolete by 2001, except as rectifiers in tube guitar and hi-fi amplifiers and in a few specialized high-voltage applications. There are two electrodes, not touching, in some kind of gas. One electrode is very sharp. The other has a smoothly curved finish. If a strong negative potential is applied to the sharp electrode, the electric field near the sharp edge or point is enough to cause an electrical discharge in the gas, and a current flows. If the reverse potential is applied, the electrical field strength around the smooth electrode is not enough to start a discharge. (The discharge can only start easily at the negative end because electrons are much more mobile than positive ions.) These are sometimes used for high-voltage high-current rectification in power supply applications. Used as voltage-controlled capacitors. These were important in PLL (phase-locked loop) and FLL (frequency-locked loop) circuits, allowing tuning circuits, such as those in television receivers, to lock quickly, replacing older designs that took a long time to warm up and lock. A PLL is faster than a FLL, but prone to integer harmonic locking (if one attempts to lock to a broadband signal). They also enabled tunable oscillators in early discrete tuning of radios, where a cheap and stable, but fixed-frequency, crystal oscillator provided the reference frequency for a voltage-controlled oscillator. Other uses for semiconductor diodes include sensing temperature, and computing analog logarithms.

REFERENCES

Bate, R., Frazier, G., Frensley, W., Reed., M., "An Overview of Nanoelectronics," *Texas Instruments Technical Journal*, July-August 1989, pp. 13-20.

Beck, J.S., J.C. Vartuli, W.J. Roth, M.E. Leonowicz, C.T. Kresge, K.D. Schmitt, C.T.-W. Chu, D.H. Olsen, E.W. Shepard, S.B. McCullen, J.B. Higgins, and J.L. Schlenker. 1992. *J. Am. Chem. Soc.* 114:10834.

Bensaude-Vincent, B.: 2001, 'The Construction of a Discipline: Materials Science in the United States', *Historical Studies in the Physical and Biological Sciences*, 31, 223-248.

Bensaude-Vincent, B.: 2004, 'Two Cultures of Nanotechnology?', *Hyle: International Journal for Philosophy of Chemistry*, 10(2), 65-82.

Berube, D.M.: 2004, 'The Rhetoric of Nanotechnology', in: D. Baird, A. Nordmann & J. Schummer (eds.), *Discovering the Nanoscale*, Amsterdam: IOS Press, pp. 173-192.

Bowes, C.L., A. Malek, and G.A. Ozin. 1996. *Chem. Vap. Deposition* 2:97.

Braun, P.V., P. Osenar, and S.I. Stupp. 1996. *Nature*. 368: 2.

Bueno, O.: 2004, 'Von Neumann, Self-Reproduction and the Constitution of Nanophenomena', in: D. Baird, A. Nordmann & J. Schummer (eds.), *Discovering the Nanoscale*, Amsterdam: IOS Press, pp. 101-115.

Carroll, J.S.: 2001, 'Social Science Research Methods for Assessing Societal Implications of Nanotechnology', in: M.C. Roco and W.S. Bainbridge (eds.), *Societal Implications of Nanoscience and Nanotechnology*, Dordrecht: Kluwer, pp. 188-192.

Chen, C-Y., S.L. Burkett, H.-X. Li, and M.E. Davis. 1993. *Microporous Mater*. 2:27.

Chianelli, R.R. 1998. Synthesis, fundamental properties and applications of nanocrystals, sheets, and fullerenes based on layered transition metal chalcogenides. In *R&D status and trends,* ed. Siegel et al.

Dresselhaus, M.S., G. Dresselhaus, and P. Eklund. 1996. *Science of fullerenes and carbon nanotubes*. San Diego: Academic Press.

Drexler, K.E.: 2004, 'Nanotechnology: From Feynman to Funding', *Bulletin of Science, Technology & Society*, 24, 21-27.

Dupuy, J.-P.: 2004, 'Complexity and Uncertainty: A Prudential Approach To Nanotechnology', in: European Commission (Community Health and Consumer Protection): *Nanotechnologies: A Preliminary Risk Analysis on the Basis of a Workshop, Brussels, 1-2 March 2004*, pp. 71-93 (www.europa.eu.int/comm/health/ ph_risk/documents/ev_20040301_en.pdf).

Dupuy, J.-P.: 2004, 'Pour une évaluation normative du programme nanotechnologique', *Annales des Mines*, (February 2004), 27-32.

Einsiedel, E.F.; Goldenberg, L.: 2004, 'Dwarfing the Social? Nanotechnology Lessons from the Biotechnology Front', *Bulletin of Science, Technology & Society*, 24, 28-33.

ETC: 2003a, *The Big Down: Atom Tech ' Technologies Converging at the Atomic Scale*. Winnipeg, Canada: Action Group on Erosion, Technology and Concentration [www.etcgroup.org/ documents/TheBigDown.pdf].

ETC: 2003b, 'No Small Matter II: The Case for a Global Moratorium ' Size Matters!' *Occasional Paper Series* 7(1) [www.etcgroup.org/documents/Occ.Paper_Nanosafety.pdf].

Etzkowitz, H., 2001, 'Nano-Science and Society: Finding a Social Basis for Science Policy', in: M.C. Roco and W.S. Bainbridge (eds.), *Societal Implications of Nanoscience and Nanotechnology*, Dordrecht: Kluwer, pp. 121-128.

European Comission: *European Workshop on Social and Economic Research on Nanotechnologies and Nanosciences, Brussels, 14-15 April 2004* [www.stage-research.net/STAGE/PAGES/ Nano.html].

European Commission: 2004, *Converging Technologies: Shaping the Future of European Societies (Report of the High Level Expert Group "Foresighting the New Technology Wave")*, Brussels: European Commission Research [http://europa.eu.int/comm/research/conferences/2004/ntw/index_en.html].

F. Buot, "Mesoscopic Physics and Nanoelectronics: Nanoscience and Nanotechnology," *Physics Reports*, pp.73-174, 1993.

F. Capasso and S. Datta, "Quantum Electron Devices," *Physics Today*, pp.74-82, February 1990.

Feynman, R., "There's Plenty of Room at the Bottom: An invitation to Enter a New Field of Physics," Talk at the Annual Meeting of the American Physical Society, 29 December 1959. Reprinted in Appendix B of Crandall and Lewis. See citation above.

Fielder, F.A.; Reynolds, G.H.: 1994, 'Legal Problems of Nanotechnology: An Overview', *Southern California Interdisciplinary Law Journal*, 3, 593-629.

Fleischer, Th.; Decker, M. & Fiedeler, U. (eds.): 2004, *Große Aufmerksamkeit für kleine Welten ' Nanotechnologie und ihre Folgen*, special issue of *Technikfolgenabschätzung ' Theorie und Praxis*, 13 (2), 5-85 [www.itas.fzk.de/tatup/042/inhalt.htm].

Fogelberg, H. & Glimell, H.: 2003, 'Molecular Matters: In Search of the Real Stuff', in: H. Fogelberg & H. Glimell, *Bringing Visibility to the Invisible: Towards A Social Understanding of Nanotechnology*, Göteborg: Göteborg University, pp. 5-32 [www.sts.gu.se/publications/STS_report_6.pdf].

Fogelberg, H. & Glimell, H.: 2003, *Bringing Visibility to the Invisible: Towards A Social Understanding of Nanotechnology*, Göteborg: Göteborg University [www.sts.gu.se/publications/STS_report_6.pdf].

Fogelberg, H.: 2003, 'The Grand Politics of Technoscience: Contextualizing Nanotechnology', in: H. Fogelberg & H. Glimell, *Bringing Visibility to the Invisible: Towards A Social Understanding of Nanotechnology*, Göteborg: Göteborg University, pp. 33-54 [www.sts.gu.se/publications/STS_report_6.pdf].

Fogelberg, H.: 2003, 'The Material Culture of Nanotechnology', in: H. Fogelberg & H. Glimell, *Bringing Visibility to the Invisible: Towards A Social Understanding of Nanotechnology*, Göteborg: Göteborg University, pp. 99-114.

Frazier, G., "An Ideology For Nanoelectronics," in *Concurrent Computations: Algorithms, Architecture, and Technology*, Plenum Press, New York, 1988.

Glimell, H.: 2001, 'Challenging Limits ' Excerpts from an Emerging Ethnography of Nano Physicists', in: H. Glimell & O. Johlin (eds.), *The Social Production of Technology: On the Everyday Life with Things*, Göteburg, SE: BAS Publisher, chapter 7, pp. 111-131 (reprinted in: H. Fogelberg & H. Glimell, *Bringing Visibility to the Invisible: Towards A Social Understanding of Nanotechnology*, Göteborg: Göteborg University, pp. 115-139.

Glimell, H.: 2001, 'Dynamics of the Emerging Field of Nanoscience', in: M.C. Roco and W.S. Bainbridge (eds.), *Societal Implications of Nanoscience and Nanotechnology*, Dordrecht: Kluwer, pp. 156-160.

Glimell, H.: 2003, 'A Nano Narrative: The Mircopolitics of a New Generic Enabling Technology', in: H. Fogelberg & H. Glimell, *Bringing Visibility to the Invisible: Towards A Social Understanding of Nanotechnology*, Göteborg: Göteborg University, pp. 55-77 [www.sts.gu.se/publications/STS_report_6.pdf].

Glimell, H.: 2003, 'Dynamics of the Emerging Field of Nanoscience', in: H. Fogelberg & H. Glimell, *Bringing Visibility to the Invisible: Towards A Social Understanding of Nanotechnology*, Göteborg: Göteborg University, pp. 79-85 [www.sts.gu.se/publications/STS_report_6.pdf].

Glimell, H.: 2004, 'Grand Visions and Lilliput Politics: Staging the Exploration of the 'Endless Frontier", in: D. Baird, A. Nordmann & J. Schummer (eds.), *Discovering the Nanoscale*, Amsterdam: IOS Press, pp. 231-246.

Goldhaber-Gordon, D., Montemerlo, M.S., Love, J.C., Opiteck, G.J., and Ellenbogen, J.C., "Overview of Nanoelectronic Devices," submitted to the *Proceedings of the IEEE*, February 1997. For more information, please send e-mail to nanotech@mitre.org.

Gorman, M.E.; Groves, J.F. & Shrager, J.: 2004, 'Societal Dimensions of Nanotechnology as a Trading Zone: Results from a Pilot Project', in: D. Baird, A. Nordmann & J. Schummer (eds.), *Discovering the Nanoscale*, Amsterdam: IOS Press, pp. 63-73.

Grinbaum, A. & Dupuy, J.-P.: 2004, 'Living with Uncertainty: Toward the Ongoing Normative Assessment of Nanotechnology', *Techne: Research in Philosophy and Technology*, 8(3) (forthcoming).

Grinbaum, A.: 2004, 'La condition de l'homme moderne et les nanotechnologies', in: G. Nivat (ed.), *Les limites de l'humain. 39èmes Rencontres Internationales de Genève*, L'Age d'Homme, Genève, p. 141.

Grunwald, A.: 2004, 'Ethische Aspekte der Nanotechnologie. Eine Felderkundung', *Technikfolgenabschätzung ' Theorie und Praxis*, 13 (2), 71-78.

Gupta, V.K. & Pangannaya, N.B.: 2000, 'Carbon nanotubes: bibliometric analysis of patents', *World Patent Information*, 22, 185-189.

Hansson, S.O.: 2004, 'Great Uncertainty about Small Things', *Techne: Research in Philosophy and Technology*, 8(3) (forthcoming).

Haruta, M. 1997. *Catalyst surveys of Japan* 1:61 and references therein.

Hennig, J.: 2004, 'Changes in the Design of Scanning Tunneling Microscopic Images from 1980 to 1990', *Techne: Research in Philosophy and Technology*, 8(3) (forthcoming).

Hessenbruch, A.: 2004, 'Nanotechnology and the Negotiation of Novelty', in: D. Baird, A. Nordmann & J. Schummer (eds.), *Discovering the Nanoscale*, Amsterdam: IOS Press, pp. 135-144.

Hullmann, A. & Meyer, M.: 2003, 'Publications and Patents in Nanotechnology. An overview of previous studies and the state of the art', *Scientometrics*, 58 (3) 507-527.

Huo, Q., D.I. Margolese, U. Ciesla, P. Feng, T.E. Gier, P. Sieger, R. Leon, P.M. Petroff, F. Schuth, and G.D. Stucky. 1994. *Nature* 368:317.

Jena, P., S.N. Khanna, and B.K. Rao. 1996. In *Science and technology of atomically engineered materials*, ed. P. Jena. River Edge, NJ: World Scientific.

Johansson, M.: 2003, 'Plenty of room at the bottom: Towards an anthropology of nanoscience', *Anthropology Today*, 19 (No. 6), 3-6.

Johnson, A.: 2004, 'The End of Pure Science: Science Policy from Bayh-Dole to the NNI', in: D. Baird, A. Nordmann & J. Schummer (eds.), *Discovering the Nanoscale*, Amsterdam: IOS Press, pp. 217-230.

Jounet, C., W.K. Maser, P. Bernier, A. Loiseau, M. Lamy de la Chapelle, S. Lefrant, P. Deniard, R. Lee, and J.E. Fischer. 1997. *Nature* 388:756.

K.E. Drexler, *Nanosystems: Molecular Machinery, Manufacturing, and Computation,* Wiley, New York, 1992.

Karger, J., and D.M. Ruthven. 1992. *Diffusion in zeolites*. New York: J. Wiley.

Keiper, A.: 2003, 'The Nanotechnology Revolution', *The New Atlantis*, 2, 17-34.

Khushf, G.: 2004, 'A Hierarchical Architecture for Nano-scale Science and Technology: Taking Stock of the Claims About Science Made By Advocates of NBIC Convergence', in: D. Baird, A. Nordmann & J. Schummer (eds.), *Discovering the Nanoscale*, Amsterdam: IOS Press, pp. 21-33.

Khushf, G.: 2004, 'Systems Theory and the Ethics of Human Enhancement: A Framework for NBIC Convergence', *Annals of the New York Academy of Sciences*, 1013, 124-149.

Krätschmer, W., L.D. Lamb, K. Fostiropoulos, and D.R. Huffman. 1990. *Nature* 347:354.

Kresge, C.T., M.E. Leonowicz, W.J. Roth, J.C. Vartuli, and J.S. Beck. 1992. *Nature* 359:710.

Kupperman, A., S. Nadimi, S. Oliver, G. Ozin, J. Garcés, and M. Olken. 1993. *Nature* 365:239.

Kuusi, O.; Meyer, M.: 2002, 'Technological generalizations and leitbilder ' the anticipation of technological opportunities', *Technological Forecasting & Social Change*, 69, 625-639.

Landon, B.: 2004, 'Less is More: Much Less is Much More: The Insistent Allure of Nanotechnology Narratives in Science Fiction', in: N.K. Hayles (ed.), *Nanoculture: Implications of the New Technoscience*, Bristol, UK: Intellect Books, pp. 131-146.

Laszlo, P.: 2004, 'Is There Life After Partington?', *Hyle: International Journal for Philosophy of Chemistry*, 10(2), 169-178.

Lenhard, J.: 2004, 'Nanoscience and the Janus-Faced Character of Simulations', in: D. Baird, A. Nordmann & J. Schummer (eds.), *Discovering the Nanoscale*, Amsterdam: IOS Press, pp. 93-100.

Lent, C.S., Tougaw, P.D., Porod, W., Bernstein, G.H., "Quantum Cellular Automata," *Nanotechnology*, Vol. 4, p. 49, 1993.

Lewak, S.E.: 2004, 'What's the Buzz? Tell Me What's A-Happening: Wonder, Nanotechnology, and Alice's Adventures in Wonderland', in: N.K. Hayles (ed.), *Nanoculture: Implications of the New Technoscience*, Bristol, UK: Intellect Books, pp. 201-.

Lide, D.R., ed. 1993-1994. *CRC Handbook of Chemistry and Physics*, 74th ed.

Lin-Easton, P.C.: 2001, 'It's Time for Environmentalists to Think Small ' Real Small: A Call for the Involvement of Environmental Lawyers in Developing Precautionary Policies Molecular Nanotechnology', *Georgetown International Law Review*, 14, 106-134.

López, J.: 2004, 'Bridging the Gaps: Science Fiction in Nanotechnology', *Hyle: International Journal for Philosophy of Chemistry*, 10(2), 129-152.

Lösch, A.: 2004, 'Nanomedicine and Space: Discursive Orders of Mediating Innovations', in: D. Baird, A. Nordmann & J. Schummer (eds.), *Discovering the Nanoscale*, Amsterdam: IOS Press, pp. 193-202.

Marshall, K.: 2004, 'Atomizing the Risk Technology', in: N.K. Hayles (ed.), *Nanoculture: Implications of the New Technoscience*, Bristol, UK: Intellect Books, pp. 147-160.

Martin, T.P., N. Malinowski, U. Zimmerman, U. Naher, and H. Schaber. 1993. *J. Chem. Phys.* 99:4210.

Martin, T.P., U. Naher, H. Schaber, U. Zimmerman. 1993. *Phys. Rev. Lett.* 70:3079.

Mayer, S.: 2002, 'From genetic modification to nanotechnology: the dangers of 'sound science", in: T. Gilland (ed.), *Science: Can We Trust the Experts?*, London: Hodder and Stoughton, pp. 1-15.

Mehta, M.D.: 2002, 'Nanoscience and Nanotechnology: Assessing the Nature of Innovation in These Fields', *Bulletin of Science, Technology & Society*, 22(4), 269-273.

Mehta, M.D.: 2004, 'From Biotechnology to Nanotechnology: What Can We Learn from Earlier Technologies?, *Bulletin of Science, Technology & Society*, 24, 34-39.

Meyer, M. & Kuusi, O.: 2004, 'Nanotechnology: Generalizations in an Interdisciplinary Field of Science and Technology', *Hyle: International Journal for Philosophy of Chemistry*, 10(2), 153-168.

Meyer, M.: 2000a, 'Does science push technology? Patents citing scientific literature', *Research Policy*, 29, 409-434.

Meyer, M.: 2000b, 'Patent citations in a novel field of technology: What can they tell about interactions of emerging communities of science and technology?', *Scientometrics*, 48, 151-178.

Meyer, M.: 2001a, 'Patent citations in a novel field of technology: An exploration of nano-science and nano-technology', *Scientometrics*, 51, 163-183.

Meyer, M.: 2001b, 'Socio-economic Research on Nanoscale Science and Technology: A European Overview and Illustration', in: M.C. Roco and W.S. Bainbridge (eds.), *Societal Implications of Nanoscience and Nanotechnology*, Dordrecht: Kluwer, pp. 217-241.

Milburn, C.: 2002, 'Nanotechnology in the age of post-human engineering: science fiction as science', *Configurations*, 10, 261-295 [reprinted in: N.K. Hayles (ed.), *Nanoculture: Implications of the New Technoscience*, Bristol, UK: Intellect Books, 2004, pp. 109-130].

Milburn, C.: 2004, 'Nano/Splatter: Disintegrating the Postbiological Body', *New Literary History*, 35 (in print).

Mnyusiwalla, A.; Abdallah, S.D.; Singer, P.A.: 2003, 'Mind the gap: science and ethics in nanotechnology', *Nanotechnology*, 14, R9-R13.

Mody, C.C.M.: 2004, 'How Probe Microscopists Became Nanotechnologists', in: D. Baird, A. Nordmann & J. Schummer (eds.), *Discovering the Nanoscale*, Amsterdam: IOS Press, pp. 119-133.

Mody, C.C.M.: 2004, 'Instruments in Training: The Growth of American Probe Microscopy in the 1980s', in: D. Kaiser (ed.), *Pedagogy and the Practice of Science: Producing Physical Scientists, 1800-2000*, Cambridge, MA: MIT Press (forthcoming).

Mody, C.C.M.: 2004, 'Small, but Determined: Technological Determinism in Nanoscience', *Hyle: International Journal for Philosophy of Chemistry*, 10(2), 99-128.

Mody, C.C.M.: 2004, *Crafting the Tools of Knowledge: The Invention, Spread, and Commercialization of Probe Microscopy, 1960-2000*, Ph.D. dissertation, Cornell University.

Montemerlo, M.S., Love, J.C., Opiteck, G.J., Goldhaber, D. J., and Ellenbogen, J.C., "Technologies and Designs for Electronic Nanocomputers," MITRE Technical Report 96W0000044, The MITRE Corporation, McLean, VA, July 1996. For more information, send e-mail to nanotech@mitre.org.

Moor, J.H. & Weckert, J.: 2004, 'Nanoethics: Assessing the Nanoscale From an Ethical Point of

View', in: D. Baird, A. Nordmann & J. Schummer (eds.), *Discovering the Nanoscale*, Amsterdam: IOS Press, pp. 301-310.

Munn Sanchez, E.: 2004, 'The Expert's Role in Nanoscience and Technology', in: D. Baird, A. Nordmann & J. Schummer (eds.), *Discovering the Nanoscale*, Amsterdam: IOS Press, pp. 257-266.

Nordmann, A.: 2003, 'Shaping the World Atom by Atom: Eine nanowissenschaftliche WeltBildanalyse', in: A. Grunwald (ed.), *Technikgestaltung zwischen Wunsch und Wirklichkeit*, Berlin: Springer, pp. 191-199.

Nordmann, A.: 2004, 'Molecular Disjunctions: Staking Claims at the Nanoscale', in: D. Baird, A. Nordmann & J. Schummer (eds.), *Discovering the Nanoscale*, Amsterdam: IOS Press, pp. 51-62.

Nordmann, A.: 2004, 'Nanotechnology: Convergence and Integration', Presentation at *EuroNanoForum, Trieste, December 10, 2003* (Proceedings in preparation).

Nordmann, A.: 2004, 'Nanotechnology's WorldView: New Space for Old Cosmologies', *IEEE Technology and Society Magazine*, 23 (forthcoming).

Nordmann, A.: 2004, 'Social Imagination for Nanotechnology', in: European Commission (Community Health and Consumer Protection): *Nanotechnologies: A Preliminary Risk Analysis on the Basis of a Workshop, Brussels, 1-2 March 2004*, pp. 111-113 [www.europa.eu.int/comm/health/ph_risk/documents/ev_20040301_en.pdf].

Nordmann, A.: 2004, 'Was ist TechnoWissenschaft? ' Zum Wandel der Wissenschaftskultur am Beispiel von Nanoforschung und Bionik', in: T. Rossmann & C. Tropea (eds.), *Bionik ' Neue Forschungsergebnisse aus Natur-, Ingenieur- und Geisteswissenschaften*, Berlin: Springer, 2004 (forthcoming)

Paschen, H.; Coenen, C.; Fleischer, T.; Grünwald, R.; Oertel, D. & Revermann, C.: 2004, *Nanotechnologie: Forschung, Entwicklung, Anwendung*, Berlin: Springer (*Nanotechnologie, TAB-Arbeitsbericht 92*, Berlin: Büro für Technikfolgen-Abschätzung beim Deutschen Bundestag).

Pitt, J.C.: 2004, 'The Epistemology of the Very Small', in: D. Baird, A. Nordmann & J. Schummer (eds.), *Discovering the Nanoscale*, Amsterdam: IOS Press, pp. 157-163.

Pressman, J.: 2004, 'Nano Narrative: A Parable from Electronic Literature', in: N.K. Hayles (ed.), *Nanoculture: Implications of the New Technoscience*, Bristol, UK: Intellect Books, pp. 191-200.

Prigogine, I., and S. Rice. 1988. *Advances in chemical physics*, Vol. 70, Parts 1 & 2. New York: J. Wiley.

R. T. Bate, "Nanoelectronics," *Nanotechnology*, Vol. 1, pp. 1-7, 1990.

Rademann, K., B. Kaiser, U. Even, F. Hensel. 1987. *Phys. Rev. Lett.* 59:2319.

Rao, C.N.R., B.C. Satishkumar, and A. Govindaraj. 1997. *Chem. Commun.* 1581.

Rao, M.B., and S. Sircar. 1993. *Gas Separation and Purification* 7:279.

Reed, M.A., "Quantum Dots," *Scientific American,* January 1993, pp. 118-123.

Roberts, J.A.: 2004, 'Deciding the Future of Nanotechnologies: Legal Perspectives on Issues of Democracy and Technology', in: D. Baird, A. Nordmann & J. Schummer (eds.), *Discovering the Nanoscale*, Amsterdam: IOS Press, pp. 247-255.

Robinson, C.: 2004, 'Images in NanoScience/Technology', in: D. Baird, A. Nordmann & J. Schummer (eds.), *Discovering the Nanoscale*, Amsterdam: IOS Press, pp. 165-169.

Robison, W.L.: 2004, 'Nano-Ethics', in: D. Baird, A. Nordmann & J. Schummer (eds.), *Discovering the Nanoscale*, Amsterdam: IOS Press, pp. 285-300.

Roco, M.C. & Bainbridge, W.S. (eds.): 2001, *Societal implications of nanoscience and nanotechnology*, (Proceedings of a workshop organized by the National Science Foundation, September 28-29, 2000), Kluwer: Dordrecht [available online at http://itri.loyola.edu/nano/societalimpact/nanosi.pdf]

Roco, M.C. & Tomellini, R. (eds.): 2002, *Nanotechnology: Revolutionary Opportunities and Societal Implications, Workshop, Lecce (Italy), 31 January - 1 February 2002*, Luxemburg: Office for Official Publications of the European Communities, [200 pp.]

Roco, M.C.; Bainbridge, W.S. (eds.): 2002, *Converging Technologies for Improving Human Performance: Nanotechnology, Biotechnology, Information Technology and the Cognitive Science*, Arlington, VA: National Science Foundation.

Roher, H. 1993. *Jpn. J. Appl. Phys.* 32:1335.

Rohlfing, E.A., D.M. Cox, and A. Kaldor. 1984. *J. Chem. Phys.* 81:3846.

Rosen, A. 1998. A periodic table in three dimensions: A sightseeing tour in the nanometer world. In *Advances in quantum chemistry*. In press.

Ruthven, D.M., S. Farooq, K.S. Knaebel. 1994. *Pressure swing adsorption*. New York: VCH Publishers.

Sarewitz, D.; Woodhouse, E.: 2003, 'Small is Powerful', in: A. Lightman, D. Sarewitz & Chr. Desser, (eds.), *Living with the Genie: Essays on Technology and the Quest for Human Mastery*, Washington, DC: Island Press, pp. 63-83.

Schiemann, G.: 2004, 'Dissolution of the Nature-Technology Dichotomy? Perspectives on Nanotechnology from an Everyday Understanding of Nature', in: D. Baird, A. Nordmann & J. Schummer (eds.), *Discovering the Nanoscale*, Amsterdam: IOS Press, pp. 209-213.

Schmidt, J.C.: 2004, 'Unbounded Technologies: Working Through Technological Reductionism of Nanotechnology', in: D. Baird, A. Nordmann & J. Schummer (eds.), *Discovering the Nanoscale*, Amsterdam: IOS Press, pp. 35-50.

Schummer, J.: 2004, 'Why do Chemists Perform Experiments?', in: D. Sobczynska, P. Zeidler, E. Zielonacka-Lis (eds.), *Chemistry in the Philosophical Melting Pot*, Peter Lang (Frankfurt/M.), 2004, pp. 395-410.

Schummer, J.: 2005, 'Reading Nano: The Public Interest in Nanotechnology as Reflected in Book Purchase Patterns", *Public Understanding of Science*, 14 (2) (forthcoming).

Schummer, J.; Baird, D. (eds.): 2004-5, *Nanotech Challenges*, joint special issue of *Hyle: International Journal for Philosophy of Chemistry & Techne: Research in Philosophy and Technology* (forthcoming).

Stuckless, J.T, D.E. Starr, D.J. Bald, C.T. Campbell. 1997. *J. Chem. Phys.* 107:5547.

Suchman, M.: 2001, 'Envisioning Life on the Nano-Frontier', in: M.C. Roco and W.S. Bainbridge (eds.), *Societal Implications of Nanoscience and Nanotechnology*, Dordrecht: Kluwer, pp. 211-216.

Suchman, M.: 2002, 'Social Science and Nanotechnologies', M. Roco & R. Tomellini (eds.),

Nanotechnology ' Revolutionary Opportunities and Societal Implications, Luxembourg: European Communities, pp. 95-99.

Sweeney, A.E.; Seal, S. & Vaidyanathan, P.: 2003, 'The promises and perils of nanoscience and nanotechnology: Exploring emerging social and ethical issues', *Bulletin of Science, Technology & Society*, 23 (4), 236-245.

Verbeek, A. & Callaert, J.: 2002, *Detailed analysis of the science-technology interaction in the field of Nanotechnology*, Luxemburg: Office for Official Publications of the European Communities (Linking Science to Technology. Bibliographic References in Patents, Vol. 9).

2

Nanoengineering, Self-assembly Nanoscale: Opportunities & Challenges

Opportunities in Self-assembly and Nano-engineering

In self-assembly large molecular structures are obtained from the organization of a large number of molecules or atoms into a given shape, typically through specific interactions of the molecules among themselves and with a template. The interaction of the different bonding mechanisms is an area of strong fundamental research interest. Only two areas will be highlighted here: zeolites and carbon materials. Both of these materials exhibit characteristics of self-assembly, namely novel and reproducible structures that can be fabricated in industrially significant quantities.

Zeolitic Materials

Aluminosilicates (e.g., zeolites) are crystalline porous nanostructures with long range crystalline order with pore sizes which can be varied from about 4 Å to 15 Å in conventional zeolites. A 3-dimensional (e.g., MFI) zeolite cage structure together with a depiction of the straight and ziz-zag channels and a 2-dimensional zeolite with channels only in 2 directions. The vertices in the stick drawing denote position of the O atoms in the crystalline lattice. This particular zeolite has 10 atoms in the zeolite "window." The size of the window is determined by the number of oxygens in the ring. Typical zeolite structures together depicting the positions of the O atoms and two different zeolitic structures one (lower left) with a three dimensional structure and (lower right) a zeolite with a two dimensional channel structure. For example normal hexane with a kinetic molecular diameter of about 5.1 Å can pass through a 10 ring or larger, whereas cyclohexane with a kinetic molecular diameter of 6.9 Å would be hard pressed to pass through a 10 ring. Thus all other things being equal, a 10-ring zeolite could be used to separate mixtures of normal hexane and cyclohexane. It is this property together with the ability to chemically modify the acidity of zeolitic materials that makes them extremely valuable as selective sorbants, as membranes and for use in selective catalytic reactions. In 1992, a new family of aluminosilicates (M41S) with pores sizes between 20 and 100 Å in diameter were reported by Mobil researchers (Beck et al. 1992; Kresge et al. 1992). One of particular interest is MCM-41, which consists of hexagonal arrays of uniform 2 to 10 nanometer-sized cylindrical pores. Not only can such materials be synthesized, but novel structures such as "tubules-within-a-tubule" have been fabricated as mesoporous molecular sieves in MCM-41 (Lin and Mou 1996). Of particular interest is the possibility of expanding the so-called "liquid crystal templating" mechanism (Chen et al. 1993) to non-aluminum dopants within the silicate MCM-41 framework (Tanev et al. 1994) and to derive nonsiliceous MCM-41 type of materials (Braun et al. 1996). Another approach to synthesizing large pore and large single crystals of zeolytic materials is being pioneered by Geoffrey Ozin and his group at the University of Toronto, who have demonstrated that crystals as large as 5 mm can be synthesized (Kupperman et al. 1993). The ability to synthesize such large crystals has important implications for discovery of new sensors (selective chemical adsorbants) and membrane devices (selective transport of molecular species), since large single crystals can now be available to the laboratory researcher to carry out fundamental studies of adsorption and diffusion properties with such materials. These materials are expected to create new opportunities for applications in the fields of separations science, for use directly as molecular sieves or as new molecular sieving sorbant materials; in catalysis, as heterogeneous catalysts; and as supports for other catalytic materials as well as other novel applications (Bowes et al. 1996; Brinker 1996; Sayari 1996). The ability to synthesize zeolitic materials of precise pore size in the range between 4 and 100 Å continues to expand the possibilities for research and technological innovation in the catalytic, separations, and sorption technologies (Ruthven et al. 1994; Karger and Ruthven 1992).

Carbon Materials

The carbon-based materials of interest from a molecular self-assembly point of view include fullerenes and their relatives, including endohedral fullerenes and metal-coated fullerenes, carbon nanotubes, carbon nanoparticles, and porous carbons. Since 1990 with the discovery of techniques to produce soluble carbon in a bottle (for examples, see Krätschmer et al. 1990 and references therein), research on and with carbon materials has skyrocketed (Dresselhaus et al. 1996; Dresselhaus and Dresselhaus 1995). Not only can the molecular forms of carbon (the fullerenes and their derivatives) be synthesized, characterized, and studied for applications, but many other new carbon materials such as multi- and single-walled carbon nanotubes can now be produced in macroscopic quantities. The broad variety of carbon nanotube structures whose properties are now being examined both theoretically and experimentally. A rich literature on these new carbon materials now exists. This report will only attempt to highlight a few important recent examples in the area of high surface area materials. Of particular interest for future catalytic applications is the recent report that not only can C60 be coated with metal atoms, but that the metal coating can consist of a precise number of metal atoms. For example, C60Li12 and C60Ca32 have been identified mass spectroscopically (Martin, Malinowski, et al. 1993; Martin, Naher, et al. 1993; Zimmerman et al. 1995). C60 has been coated with a variety of different metals, including Li, Ca, Sr, Ba, V, Ta and other transition metals. Interestingly, addition of more than 3 Ta atoms to C60 breaks the C60 cage. Replacement of one carbon atom in C60 by a transition metal atom such as Co or Ir is being studied for possible catalytic applications. The future technological challenge will be to discover techniques to fabricate large quantities of such materials, so that such catalyst materials can be put in a bottle and not just in molecular beams.

Examples of carbon nanotube structures, including multiwalled and metal-atomfilled nanotubes. Carbon nanotubes have the interesting property that they are predicted to be either semiconducting or conducting (metallic), depending on the chirality and diameter of the nanotube. Such materials are being studied as conductive additives to plastics and for use in electrochemical applications where the uniformity of the nanotube diameter and length is not overly critical (Dresselhaus 1998). Another approach is to use the carbon nanotube as a template for a nanotube of an inorganic oxide. Hollow nanotubes of zirconia and yttria-stablilized zirconia have been prepared by coating treated carbon nanotubes with a zirconium compound and then burning out the carbon template (Rao et al. 1997). Finally, large scale production of singlewalled nanotubes has recently been demonstrated, so one may anticipate a strong upsurge in the characterization and potential usage of single-walled carbon nanotubes in the future (Jounet et al. 1997). Porous carbons are of interest as molecular sieve materials, both as sorbants and as membranes, or as nanostraws for filtration. One of the major research objectives is to develop materials or structures with exceedingly high storage capacity per unit volume and weight for gases such as H2 or CH4. H2 or CH4 could become an economic source of combustion fuel or a means to power fuel cells for ultralow-emission vehicles or for electric power generation. Microporous hollow carbon fibers have exhibited high permeance and high selectivity as hydrogen selective membranes, and development is now underway to scale up these membranes to commercial levels (Soffer et al. 1987; Jones and Koros 1994; Rao and Sircar 1993). Carbon fiber materials produced via catalytic decomposition of hydrocarbon vapors have also recently been reported to exhibit exceptionally high hydrogen adsorption capacity (Baker 1998). More mundane uses of nanotubes are as nanometer reinforcing rods in polymers or even in concrete. Incorporation of conducting carbon nanotubes in construction materials such as concrete or structural plastics opens opportunities for real time monitoring of material integrity and quality.

Microporous and Dense Ultrathin Films

Research and development of microporous thin films for use as molecular sieving membranes using inorganic crystalline materials such as zeolites or porous silica is another area of active research around the world. For molecular sieving membranes, one critical challenge rests on discovering ways to create large scale, thin, nearly defect-free membranes. One recent example is the fabrication of mesoporous conducting thin films grown from liquid crystal mixtures (Attard et al. 1997). Transmission electron microscopy (TEM) reveals an ordered array of 2.5 nm diameter cylindrical holes in a 300 nm thick Pt film. The hole diameter can be varied either by changing the chain length of the surfactant molecule or by adding an alkane to the plating solution. It is interesting that this technique produces a continuous thin film with nanoscale porosity in an electrically conducting material. Dense ultrathin films such as single monolayer films would be of significant importance in the semiconductor industry. Thin films of specialized coatings for corrosion, thermal, and/or chemical stability should be valuable for the chemical and aerospace industries. Novel chemical sensors may be anticipated through use of ultrathin films composed of specialized clusters. Typical techniques for production of thin films are physical vapor deposition, chemical vapor deposition, and Langmuir- Blodgett processes.

Nanoscale: Opportunities in Characterization and Manipulation

Over the last two decades, the development and improvement of new techniques to fabricate and characterize nanoscale materials have fueled much of the enormous growth in nanoscale science and technology, not only by making nanoscale materials relatively easily available for scientific study and characterization, but also in some instances, opening the door for large scale industrial use. For example, atomic force microscopy and scanning tunneling microscopy are two techniques that have become major workhorses for characterization of nanoscale materials. The strong upsurge in interest and funding in nanoscale materials must to some (large) degree be credited to the recent development of these two techniques. Combining Xray structure, high resolution TEM and low energy, high resolution scanning electron microscopy (SEM), researchers now have the means to physically characterize even the smallest structures in ways impossible just a few years ago. Not only can a nanostructure be precisely examined, but its electronic character can also be mapped out.

Using the scanning probe devices, scientists can both image individual atoms and molecules and also manipulate and arrange them one at a time. This atomic manipulation to build structures is just in its infancy, but it does allow one to imagine a route to the ultimate goal of atomically tailored materials, built up atom-by-atom by a robotic synthesizer. The "abacus" of C60 molecules produced at the IBM laboratories in 1997 is an excellent example of possibilities that may lie ahead for manipulation at the atomic scale. The Atom Technology project in Japan is now in its second five-year program to push the frontiers of atom manipulation closer to the commercial sector. Of all fundamental properties controlling the stability of atoms, clusters, and particles on a surface or support, knowledge of the adsorption and adhesive energies of the metal atom or particle on the solid metal or oxide surface is critical to fundamental understanding of the stability of high surface area materials for materials applications that include oxide-supported metal catalysts, bimetallic catalysts, and metal-ceramic interfaces used in microelectronics.

Knowledge of such parameters allows researchers to predict the relative strengths of the metal-metal and metal atom-support interaction energies, and to infer relative stabilities as a function of the composition and size of the metal cluster. Recently it has become possible to experimentally measure the metal atom-surface bond strength on a peratom basis using adsorption micro-calorimetry on ultrathin single crystal metal or metal oxide surfaces (Stuckless et al. 1997). The direct calorimetric measurement of metal adsorption energies developed at the University of Washington is based in part on earlier work first developed by D.A. King and colleagues at Cambridge University (Yeo et al. 1995). A technique such as this capable of probing interactions on an atom-by-atom or molecule-bymolecule basis can be thought of as another "atomic probe" that can be expected to substantially advance our database and understanding at the ultimate nanoscale for materials, that is, single atom binding energies to surfaces.

Similarly, new techniques are being developed to allow chemical and catalytic reactions to be followed in situ in real time. As an example, an infrared and nuclear magnetic resonance spectroscopic technique is being developed at the Max Planck Institute in Mülheim to monitor kinetics of CO adsorption on 1-3 nm diameter metal colloid particles (typically Pt, Rh, or Pd) in liquids and to follow in real time the way CO organizes itself on the particles while in liquid suspension. Such techniques will allow one to begin to understand the metal particle properties in solution and thus infer what might occur in real reaction mixtures. Extension of such techniques to real catalytic reactions in solution for catalyst particles of various sizes and composition is likely in the not too distant future.

The areas where nanoscale high surface area materials may have the greatest future impact are difficult to predict, but some signs point to the possibility of substantial advancement in the areas of adsorption/separations, particularly in gas sorption and separations and in novel chemical catalysis using nanoscale catalyst particles. At least two major challenges must be faced before utilization and generation of high surface area nanoscale materials becomes a commonplace reality. First is critical dimensional control of the nanoscale structure over long times and varying conditions. In nanoscale catalyst materials the critical chemical selectivity is likely to be intimately associated with the local environment around what is termed the "active" site. This suggests that the size, type, and geometry of the atoms making up the active site will play a critical role in defining the conditions under which this active site will be able to carry out its designed function. Ability to fabricate materials with "exactly" the same structure and composition at each active site has been and will continue to be a major challenge to materials and catalytic scientists. A second challenge involves thermal and chemical stability control of the fabricated nanostructure. It is generally accepted that the smaller the nanostructure (active site) the more likely it is that the structure may move, aggregate, be poisoned, decompose, or change its shape, composition or morphology upon exposure to thermal and/or chemical cycling.

Identifying windows of stable operation in which the specific structure or material will be able to retain the desired (and designed) behavior is critical for commercial applications. On the other hand, the driving force is the fact that nanostructured materials typically exhibit unique properties that are expected to open windows of opportunity previously inaccessible with existing materials.

It is important to recognize that nanoscale science and technology is not a "stand alone" field of endeavor, but rather is more of a "generic" area that is expected to have a critical impact and overlap in many areas of science and technology. The breadth of issues covered in this report can be taken as proof of this principle. The fields that fall under the "nanoscale" umbrella are many and diverse, illustrating that nanoscale science and technology is a collection of many different disciplines and areas of expertise. Such science and technology offers both an opportunity and a challenge to the scientific and technological community. Researchers in some areas of science included under the broad umbrella of nanoscale science and technology in this report do not normally consider (or in some cases, want) their scientific efforts to be labeled as "nanoscale." For this chapter, nanoscale science and technology broadly encompasses the science and technology that falls between that involving individual atoms or molecules at the one extreme and that involving "bulk" materials at the other extreme. In summary, it is important to recognize that the use of nanostructuring or nanostructures to generate, fabricate or assemble high surface area materials is at an embryonic stage. The effect of the nanostructure and our ability to measure it will be increasingly important for future progress and development of materials for the marketplace. That said, it is apparent that so-called "mature" technologies such as catalysis, coatings, separations, etc., are already being impacted. Thus, one may eagerly anticipate exciting new advances in many diverse technological areas growing from our increasing understanding of nanostructuring and nanostructured materials.

Functional Nanoscale Devices

The recent emergence of fabrication tools and techniques capable of constructing structures with dimensions ranging from 0.1 to 50 nm has opened up numerous possibilities for investigating new devices in a size domain heretofore inaccessible to experimental researchers. The WTEC nanotechnology panel reviewed research in the United States, Japan, Taiwan, and Europe to find that there is considerable nanoscience and technology activity in university, industrial, and government laboratories around the world. The insight gained from this survey suggests areas of strength and areas of possible improvement in the field. There is intense study around the world to determine the exact point in dimensional scaling where it becomes either physically unfeasible or financially impractical to continue the trend towards reducing the size while increasing the complexity of silicon chips. In some of the same laboratories where research activities on Si are decreasing, research activities on singleelectron devices (SEDs) are increasing. Although there are myriad questions involving electrical contacts, interconnections, reliability, and the like, one of the fundamental issues in the miniaturization/complexity debate concerns the Si MOSFET itself when the gate length is reduced to less than 50 nm. Does it behave like a long gate device or does the output conductance increase to impractical levels due to short-channel effects? Based on the WTEC panel's survey, most of the activities examining these questions are taking place in Japanese industrial laboratories.

Functional device scales. While the signature current-voltage (I-V) characteristics provide a common basis for comparison of device performance, there are significant variations in the fabrication methods and device structures being considered by the different labs in the countries the panel surveyed that have significant SED activity. The range of research in the surveyed laboratories spans electrical measurements from millikelvin to room temperature and from discrete electronic elements to integrated single-electron transistors (SETs). Materials that are used to form the active single-electron element range from charge clusters that are shaped by electric fields in a two-dimensional electron gas to metallic colloids to single oligomers. Progress in the field is hindered by architectures based on conventional circuit approaches that fail to take sufficient advantage of the unique properties of single-charge electronics to achieve significant impact in future high density applications. Most research in SED technology is fundamental and is distributed among universities funded by government agencies. A smaller body of applicationdirected research exists in industrial laboratories; these are mostly in Japan. The field of magnetics has experienced increasing attention since giant magnetoresistance (GMR) in multilayered structures was discovered in 1988. In these structures ferromagnetic layers are quantum mechanically coupled across a 1-3 nm nonmagnetic metallic layer. GMR structures are under intense study for applications in hard disk heads, random access memory (RAM), and sensors. Several laboratories are investigating the physics of the transition of these layers, which are quantum mechanically confined in one dimension, to layered filaments in which there are one- and two-dimensional confinements. There are numerous experimental process approaches under consideration in fabricating GMR structures, including the following:

- magnetron or ion beam sputter deposition
- epitaxy for layered structures
- rubber stamping of nanoscale wire-like patterns
- electroplating into nanoscale pores in polymer membranes

In RAM applications, a high ratio of magnetoresistance combined with a small coercive switching field is key to density, speed, and low power. These features are also achieved in magnetic tunnel junctions in which the ferromagnetic layers are quantum mechanically coupled through a thin dielectric layer. Although research in nanoscale magnetics is underway internationally, most of the activities on the practical applications mentioned above are in the United States. Optical devices have already benefited from incorporation of nanostructured materials: commercially available semiconductor lasers incorporate active regions comprised of quantum wells, the presence of which modifies the electronic density of states and the localization of electrons and holes, resulting in more efficient laser operation. Extrapolating from those results, even greater improvements are predicted for lasers utilizing either quantum wire or quantum dot active layers. Recent advances in the "self-assembled" formation of quantum dot structures have stimulated progress in the fabrication and characterization of quantum dot lasers in Japan, Europe, and the United States. In late 1991, the first synthesis and characterization of carbon nanotubes were reported. The novel material contained a wide variety of multiwalled nanotubes (MWNT) containing 2 to 50 concentric cylindrical graphene sheets with a diameter of a few nm and a length of up to 1 µm. The material was produced at the negative electrode of an arc discharge and appeared to be mixed with a large amount of other forms of carbon. This initial work led many groups throughout the world to produce and purify nanotubes. The theoretical study of their electronic structure followed in the next year.

Nanoengineering

A few months ago I was asked the question,

> "If I wanted to one day become a nano-engineer who designed nano-robots, what subjects should I study?"
>
> I thought for for a while before reluctantly answering with something like, "NEMS and and anything quantum." I knew that wasn«t the best answer, but I couldn«t think of anything else. Those just happened to be the things I thought were important while I tried to figure out *quantum biology* and which stocks fit into *The Nanotech Sector*. After a bit more thought and a literature search, I thought of a better response, *engineering*.

The objective behind nanoengineering isn't much different from ordinary engineering; that is, to design and assemble functional devices from the available components. Nanoengineers are already assembling those components into functional devices such as *nanoparticle films* for space aged coatings and *gene chips* for medical diagnostics. Furthermore, assembling molecular components into circuits for bottom-up integration is a promising route towards the fabrication of really fast *quantum computers*. With these examples as a few motivating factors, rapid progress is being made in our ability to plan and carry out molecular assembly.

A diverse array of components is available to the nanoengineer who wishes to use bottom-up design for the fabrication of devices such as a "nano-robot." Biological compounds such as nucleic acids, proteins and lipids are particularly promising components since they are already known to carry out important nanoscale functions in biological systems. Synthetic and semi-synthetic components are also useful for nanotechnology. *Quantum dots*, *nanowires* and *thin films* are a few examples.

Assembly of these components requires some technique. One way to assemble nanodevices is through *biomolecular self-assembly*. Biological compounds frequently have the innate ability to spontaneously assemble into the nanoscale machinery that makes life possible. Self-assembly of synthetic components is also common practice among nanoengineers. For instance, strategically shaped computer chips can be designed that self-assemble into functional *3D networks*. Another promising method is through the use of localized fields, for instance, those generated by *MEMS* components such as *comb drives*.

Universal Assemblers

These second-generation nanomachines—built of more than just proteins—*will do all that proteins can do, and more*. In particular, some will serve as improved devices for assembling molecular structures. Able to tolerate acid or vacuum, freezing or baking, depending on design, enzyme-like second-generation machines will be able to use as "tools" almost any of the reactive molecules used by chemists—but they will wield them with the precision of programmed machines. They will be able to bond atoms together in virtually any stable pattern, adding a few at a time to the surface of a workpiece until a complex structure is complete. Think of such nanomachines as *assemblers*. Because assemblers will let us place atoms in *almost any reasonable arrangement (as discussed in the Notes)*, they will let us build almost anything that the laws of nature allow to exist. In particular, they will let us build almost anything we can design—including more assemblers. The consequences of this will be profound, because our crude tools have let us explore only a small part of the range of possibilities that natural law permits. Assemblers will open a world of new technologies.

Advances in the technologies of medicine, space, computation, and production—and warfare—all depend on our ability to arrange atoms. With assemblers, we will be able to remake our world or destroy it. So at this point it seems wise to step back and look at the prospect as clearly as we can, so we can be sure that assemblers and nanotechnology are not a mere futurological mirage.

Disassemblers

Molecular computers will control molecular assemblers, providing the swift flow of instructions needed to direct the placement of vast numbers of atoms. Nanocomputers with molecular memory devices will also store data generated by a process that is the opposite of assembly. Assemblers will help engineers synthesize things; their relatives, *disassemblers*, will help scientists and engineers analyze things. The case for assemblers rests on the ability of enzymes and chemical reactions to form bonds, and of machines to control the process. The case for disassemblers rests on the ability of enzymes and chemical reactions to break bonds, and of machines to control the process. Enzymes, acids, oxidizers, alkali metals, *ions*, and reactive groups of atoms called *free radicals*—all can break bonds and remove groups of atoms. Because nothing is absolutely immune to corrosion, it seems that molecular tools will be able to take anything apart, a few atoms at a time. What is more, a nanomachine could (at need or convenience) apply mechanical force as well, in effect prying groups of atoms free. A nanomachine able to do this, while recording what it removes layer by layer, is *a disassembler*. Assemblers, disassemblers, and nanocomputers will work together. For example, a *nanocomputer* system will be able to direct the disassembly of an object, record its structure, and then direct the assembly of perfect copies, And this gives some hint of the power of nanotechnology.

Assemblers will take years to emerge, but their emergence seems almost inevitable: Though the path to assemblers has many steps, each step will bring the next in reach, and each will bring immediate rewards. The first steps have already been taken, under the names of "genetic engineering" and "biotechnology." Other paths to assemblers seem possible. Barring worldwide destruction or worldwide controls, the technology race will continue whether we wish it or not. And as advances in computer-aided design speed the development of molecular tools, the advance toward assemblers will quicken.

To have any hope of understanding our future, we must understand the consequences of assemblers, disassemblers, and nanocomputers. They promise to bring changes as profound as the industrial revolution, antibiotics, and nuclear weapons all rolled up in one massive breakthrough. To understand a future of such profound change, it makes sense to seek principles of change that have survived the greatest upheavals of the past. They will prove a useful guide.

Nanotechnology and Transportation

Getting around quickly requires vehicles and somewhere for them to travel. The old 1950s vision of private helicopters would be technically possible with inexpensive, high-quality manufacturing, cheap energy, and a bit of improvement in autopilots and air-traffic control—but will people really tolerate that much junk roaring across the sky? Fortunately, there is an alternative both to this and to building ever more roads.

Near the surface of the Earth, there is as much room underground as there is above it. This is usually ignored, because the room is full of dirt, rock, pressurized water, and the like. Digging is expensive. Digging long, deep tunnels is even more expensive. This expense, however, is mostly in the cost of equipment, materials, and energy. Tunneling machines are in wide use today, and molecular manufacturing can make them more efficient and less expensive. The energy to operate them will be no great problem, and smart materials can line tunnels as fast as they are dug, with little or no labor. Nanotechnology will open the low frontier. With a little care, the environmental impact of a deep tunnel can be trivial. Instead of solid rock far below the surface, there is rock with a sealed tunnel running through it. Nothing nearby need be disturbed. Tunnels avoid both the aesthetic impact of a sky full of noisy aircraft and the environmental impact of paving strips of landscape. This will make them less expensive than roads, and they can, if desired, be more common than roads in the developed world today. They will even permit faster transportation.

Japan and Germany are actively developing magnetic trains, like those in the Desert Rose Scenario. These avoid the limitations of steel wheels on steel rails by using magnetic forces to "fly" the train along a special track. Magnetic trains can reach aircraft speeds at ground level. On long runs through evacuated tunnels, they can reach spacecraft speeds, traveling global distances in an hour or so (less, if passengers are willing to tolerate substantial acceleration). Systems like this can give "taking the subway" a new meaning. Local transportation would be at fast automotive speeds, but long-distance transportation would be faster than the Concorde. With superconducting electrical systems, fast subways would be more energy efficient than today's slow mass transit.

For decades, people have proposed replacing automobiles with some form of mass-transportation system, and it seems that cost revolutions (including inexpensive tunneling) may finally make this practical. Before junking the car, though, it's worth seeing how it might be improved. Molecular manufacturing can make almost anything better. Automobiles can be made stronger and safer, lighter, higher performance, and higher efficiency, while getting excellent mileage and burning clean, inexpensive fuels, perhaps in fuel cells powering quiet electric motors. Using aerodynamic forces to hold the car to the road, there's no reason why a comfortable passenger car shouldn't be able to deliver uncomfortable, drag-racer acceleration. To imagine a cheap car built with molecular manufacturing, first imagine loading it with all the attractive features that you've ever heard proposed. This includes everything from today's self-adjusting seats and mirrors, excellent sound systems, and specially tuned steering and suspension systems, through automated navigation displays, emergency braking, and reliable super-duper airbags. Now, instead of just having the position of the seats, mirrors, and so forth adjust to a driver, as some cars do today, our smart-material car can also adjust its size, shape, and color, facing owners with choices such as, "What should our car look like for this occasion?" Those seeking an image of solid conservatism and wealth won't drive such cheap cars; they will risk their necks in a certified antique car, made from the traditional steel, paint, and rubber. If environmental regulations permit it, the car might even have a genuine gasoline-burning engine. The latter can no doubt be cleaned up by fancy nanotechnology-based emission-control systems.

Our transportion system today effectively ends in the upper atmosphere. Travel beyond still takes the form of "historic missions." There is no reason for this situation to continue for long, once molecular manufacturing becomes well established. The cost of spaceflight is high because spacecraft are huge, fragile things, made in such small numbers that they're almost hand-crafted. Molecular manufacturing will replace today's delicate monsters with rugged, mass-produced vehicles (which, with greater efficiency, needn't be so large). The vehicles will cost little, but the energy? Today, the energy cost of a ticket to orbit in an efficient vehicle would be less than one hundred dollars. Low cost vehicles and energy will drop the total cost to a fraction of this. We will know that spaceflight has become inexpensive when people see the Earth as just a small part of the world, and understand in their bones that space resources make continued exploitation of Earth's resources unnecessary. In the long run, efficient, clean, low-cost manufacturing can transform the way human beings affect the Earth by their presence. Even stay-at-home humans will be better able to heal the damage they have done.

Protein Enginnering

A 1981 paper discussed de novo protein design as part of a long-term strategy for developing complex molecular devices and systems. It presented arguments against the view that the fold-design problem is an extension of the classical (and still unsolved) fold-prediction problem (i.e., predicting folds from sequences without homologous models), a view which has discouraged efforts at design. Fold prediction is a scientific problem: it must deal with naturally evolved sequences, but natural selection's 'design goals' enforce only the physical reliability of folding — not its human predictability. This results in folds of only minimal stability. Fold design, in contrast, is an engineering problem. Protein engineers, exploiting their freedom of design, can work with sequences artificially selected for superior predictability and stability of folding. These observations indicated that "the difficulties encountered in predicting the conformations of natural proteins do not seem insurmountable obstacles to protein engineering". In accord with the implications of this argument, we have seen the successful, de novo design of a globular protein (alpha-4) [2,3] while the classical fold prediction problem remains unsolved. Likewise confirmed has been the suggestion that design can increase protein stability beyond that enforced by natural selection. In recent years, deliberate single-residue modifications have raised protein stabilities through a variety of mechanisms. Owing to design choices consistently biased toward stability, the protein alpha-4 has a stability of 22 kcal/mole, substantially greater than the 4-9 kcal/mole of typical natural proteins of similar size. Successful protein engineering marks a milestone in a research agenda leading toward capabilities of broad technological significance.

The September 2001 issue of *Scientific American* devoted six articles and a great deal of text to various perspectives on nanotechnology. The issue included various attacks on the feasibility of molecular assemblers and the work of K. Eric Drexler and his research associates. Drexler's textbook *Nanosystems* was published in 1992. It is still being used as a technical reference all over the world, and to date no significant errors have been found. As the reader can well imagine, if there were any significant errors *Scientific American* would have pointed them out. They haven't. The technical claims and conclusions of *Nanosystems* have withstood almost a decade of serious public review. Besides *Nanosystems*, there is now a very large body of technical articles, books, conferences, newsletters, and technical discussion groups that supports the feasibility of molecular machines in general and molecular assemblers in particular.

Two specific articles in the September *Scientific American* were attempts to cast doubt on the feasibility of nonbiological molecular assemblers. George Whitesides, in "The Once and Future Nanomachine," expresses concerns about many issues that have been previously addressed in the literature. He stated: "Fabrication based on the assembler is not, in my opinion, a workable strategy and thus not a concern." For commentary and references, see the article: "Many Future Nanomachines: A Rebuttal to Whiteside's Assertion That Mechanical Molecular Assemblers Are Not Workable and Not A Concern" at http://www.imm.org/SciAmDebate2/whitesides.html. While there are quite a few questionable statements in the rest of that issue, nowhere else is there a serious attempt to advance a technical argument against the feasibility of molecular nanotechnology.

How is it possible that an otherwise respectable publication would publish these attacks? None of their technical criticisms of molecular assemblers has withstood scrutiny — all have fallen by the wayside when it became obvious that they were incorrect. Some of the reasons are reviewed in "That's impossible! How good scientists reach bad conclusions"

Nanoelectronics

Experts everywhere say that nanotechnology will lead to'smarter, faster, and more efficient electronicsl but they also'admit that researchers face some challenging obstacles along the information superhighway. Particle contamination is'one of the thorniest problems'facing the electronics and computing industries. Even the tiniest particle can destroy a computer chip and cause other damage. Several University researchers are searching for ways'to detect particle contaminants and prevent their formation. A team of researchers headed by mechanical engineering professor Peter McMurry is studying'the formation and growth of contaminants in semiconductor processing equipment. Smaller contaminated particles deposit more quickly than larger ones and can destroy a number of memory chips and processors. The team's goal is'to design new processes or tools for producing contaminant-free devices. "It's important to understand how`to prevent particle deposition, and this`will require an`understanding`of how such particles`are formed and transported," McMurry says. His team has developed a device called the particle beam mass' spectrometer, the only available instrument for measuring nanoparticles'in low-pressure semiconductor processing equipment. Electrical engineering professor Steve Campbell is among the researchers who are looking'for ways to'create electronic'devices with greater memory. His studies are aimed at detecting and preventing the formation of particle contaminants'in chemical reactors. Although research in the area is'still relatively new, some early prototypes have been made, says'Campbell.

Campbell's Microtechnology Laboratory team has developed a machine to detect and measure particlesn "Now we're trying to` demonstrate that isolating`these particles`and using them to`our advantage is`feasible," he says, adding that his group is seeking funding to continue its research. "The importance of smaller and faster semiconductor devices is evident from the trends`we've all seen [toward] faster, more powerful, and less costly computersl" says McMurry. "Work that is` being done to fabricate devices at the nanoscale will ensure that this trend will continue for some time to"come." The same is true in recording media. Associate Professor Randall Victora and Professor Jack'Judy of the electrical and computer engineering department received grants'two'years ago from the National Storage Industry Consortium and Seagate Corporation to' develop superlattice magnetic recording media. The focus of their research is magnetic'recording technology for hard drivesl which are vulnerable to thermal fluctuations. These fluctuations can affect the recording process and destroy vital information stored on a disk'in the drive. "We're preventing "that from happeningl" Victora says. "These superlattices cause the disk to`have a stronger resistance to`these thermal fluctuationsn" Once developed, the technology would be used primarily by'the computing industry. Companies such as Quantum, IBM, and Seagate already have expressed interest in the group's research, Victora says. Other University projects also have implications for the electronics industry, including the study of adhesion in computer chip manufacturingl molecular electronics, and the search for an appropriate nonconductor to replace copper, which is'limited in its ability to transport information between resistors.

REFERENCES

Axelbaum, R.L. 1997. Developments in sodium/halide flame technology for the synthesis of unagglomerated non-oxide nanoparticles. In *Proc. of the Joint NSF-NIST Conference on Nanoparticles: Synthesis, Processing into Functional Nanostructures and Characterization* (May 12-13, Arlington, VA).

Beck, J.S., J.C. Vartuli, W.J. Roth, M.E. Leonowicz, C.T. Kresge, K.D. Schmitt, C.T.-W. Chu, D.H. Olsen, E.W. Shepard, S.B. McCullen, J.B. Higgins, and J.L. Schlenker. 1992. *J. Am. Chem. Soc.* 114:10834.

Becker, M.F., J.R. Brock, H. Cai, N. Chaudhary, D. Henneke, L. Hilsz, J.W. Keto, J. Lee, W.T. Nichols, and H.D. Glicksman. 1997. Nanoparticles generated by laser ablation. In *Proc. of the Joint NSF-NIST Conf. on Nanoparticles.*

Berger, R., C. Gerber, H.P. Lang, and J.K. Gimzewski. 1996. Micromechanics: a toolbox for femtoscale science: "Towards a laboratory on a tip." *Microelectronic Engineering*, 35:373-9. (International Conference on Micro- and Nanofabrication, Glasgow, U.K., 22- 25 Sept. 1996).

Berkowitz, A.E., J.R. Mitchell, M.J. Carey, A.P. Young, S. Zhang, F.E. Spada, F.T. Parker, A. Hutten, and G. Thomas. 1992. *Phys. Rev. Lett.* 68:3745.

Berndt, C.C., J. Karthikeyan, T. Chraska, and A.H. King. 1997. Plasma spray synthesis of nanozirconia powder. In *Proc. of the Joint NSF-NIST Conf. on Nanoparticles.*

Bimberg, D., N. Kirstaedter, N.N. Ledentsov, Zh.I. Alferov, P.S. Kopev, and V.M. Ustinov. 1998. InGaAs-GaAs quantum-dot lasers. *IEEE J. of Selected Topics in Quant. Electronics* 3:196-205.

Braun, P.V., P. Osenar, and S.I. Stupp. 1996. *Nature*. 368: 2. Brinker, C.J. 1996. *Curr. Opin. Solid State Mater. Sci*. 1:798. Chen, C-Y., S.L. Burkett, H.-X. Li, and M.E. Davis. 1993. *Microporous Mater*. 2:27.

Brotzman, R. 1998. Nanoparticle Dispersions. In *R&D status and trends*, ed. Siegel et al. Bu, X., P. Feng, and G.D. Stucky. 1998. Large-cage zeolite structures with multidimensional 12-ring channels. *Science* 278:2080-2085.

Brotzman, R. 1998. Nanoparticle dispersions. In *R&D status and trends*, ed. Siegel et al. Brus, L.E. 1996. Theoretical metastability of semiconductor crystallites in high pressure phases with applications to beta tin structures of silicon. *J. Am. Chem. Soc*. 118:4834-38.

Brus, L. 1996. Semiconductor colloids: Individual nanocrystals, opals and porous silicon. *Current Opinion in Colloid & Interface Science* 1:197-201. *Chemical Engineering News*. 1997. Particulate matter health studies to be reanalyzed (August 18):33.

Chance, B., Mueller, P., DeVault, D. & Powers, L. (1980) *Phys. Today* 33 (10), 32-38.

Chang, K. "Smaller Computer Chips Built Using DNA as Template." New York Times, November 21, 2003.

Chantrell. 1994. *J. Appl. Phys.* 76:6811. Erb, U., G. Palumbo, R. Zugic, and K.T. Aust. 1996. In *Processing and properties of nanocrystalline materials*, ed. C. Suryanarayana, J. Singh, and F.H. Froes. Warrendale, PA: TMS. Gertsman, V.Y., M. Hoffman, H. Gleiter and R. Birringer. 1994. *Acta Metall. Mater*. 42:3539-3544.

Chopra, N.G., R.J. Luyken, K. Cherrey, H. Crespi, M.L. Cohen, S.G. Louie, and A. Zettl. 1995. *Science* 269:966.

Clasen, R. 1990. *Int. Journal of Glass and Science Technology* 63: 291. Crandall, B.C. and J. Lewis, (eds.). 1992. *Nanotechnology: Research and perspectives*. Cambridge: MIT Press.

Czekai, D., et al. 1994. *Use of smaller milling media to prepare nanoparticulate dispersions*. U.S. Patent application. Docket 69802(02) Filed 2/25/94.

Drexler, E., Phoenix, C.: 2004, 'Self-Replication in Nanotechnology: Feasible, Potentially Safe, and Unnecessary' (unpublished paper, in preparation).

Drexler, E.: 1986, *Engines of Creation: The Coming Era of Nanotechnology*, Anchor Books, New York (expanded edition with a new afterword, 1990).

Drexler, K.E.: 1992, *Nanosystems. Molecular machinery, manufacturing and computation*, John Wiley & Sons, New York.

Drexler, K.E.: 2001, 'Machine-Phase nanotechnology', *Scientific American*, (Sept.), 66-67.

Drexler, K.E.: 2003a, 'Open Letter', *Chemical & Engineering News*, 81, no. 48, 37-42.

Drexler, K.E.: 2003b, 'An open letter to Richard Smalley', April 16.

Dupuy, J.P. "Two Temporalities, Two Rationalities: A New Look at Newcomb's Paradox." in Bourgine, P.&Walliser, B. (eds.), Economics and Cognitive Science. New York: Pergamon. 1992, 191-220.

Fennema, O. R. (1973) in *Low-Temperature Preservation of Foods and Living Matter*, eds. Fennema, O. R., Powrie, W. D. & Marth, E. H. (Dekker, New York), pp. 436-475.

Feyerabend, P.: 1981, *Realism, Rationalism and Scientific Method* (Philosophical Papers, volume 1), Cambridge University Press, Cambridge.

Feynman R. "There's Plenty of Room At the Bottom." Talk given on at the annual meeting of the American Physical Society at the California Institute of Technology, 1959.

Feynman, R.: 1960, 'There's Plenty of Room at the Bottom', *Engineering and Science*, 23, 22-36.

Fleming, D. "The Economics of Taking Care: An Evaluation of the Precautionary Principle." in Freestone, D.&Hey, E. (eds.), The Precautionary Principle and International Law. La Haye: Kluwer Law International, 1996.

Foerster, H.v.: 1962, 'Bio-Logic', in: E.E. Bernard & M.R. Kare (eds.), *Biological prototypes and synthetic systems*, Plenum Press, New York, vol. 1., pp. 1-12

Friedlander, S.K. 1993. Controlled synthesis of nanosized particles by aerosol processes. *Aerosol Sci. Technol*. 19:527.

Friedlander, S.K. 1998. Synthesis of nanoparticles and their agglomerates: Aerosol reactors. In *R&D status and trends*, ed. Siegel et al.

Friedlander, S.K., H.D. Jang, and K.H. Ryu. 1998. *Appl. Phy. Lett*. 72(2):173.

Goddard, W.A. 1998. Nanoscale theory and simulation. In *R&D status and trends*, ed. R. Siegel et al.

Godsell, D.: 2003, *Living Machinery. Bionanotechnology: Lessons from Nature*, Wiley-Liss, New-York.

Guarnieri, F., M. Fliss, and C. Bancroft. 1996. Making DNA Add. *Science* 273:220-223. Hanes, J., J.L. Cleland, and R. Langer. 1997.

Held, R., T. Heinzel, A.P. Studerus, K. Ensslin, and M. Holland. 1997. Semiconductor quantum point contact fabricated by lithography with an atomic force microscope. *Appl. Phys. Lett.* 71:2689-91.

Hellemans, A. 1998. X-rays find new ways to shine. *Science* 277:1214-15.

Hoch, H.C., L.W. Jelinski, and H.G. Craighead, eds. 1996. *Nanofabrication and biosystems*. New York: Cambridge University Press.

Howland, H.: 1962, Structural, hydraulic and economic aspects of leaf venation and shape', in: E.E. Bernard & M.R. Kare (eds.), *Biological prototypes and synthetic systems*, Plenum Press, New York, vol. 1, pp. 183-192 .

Hoyningen-Huene, P.: 1993, *Reconstructing Scientific Revolutions: Thomas S. Kuhn's Philosophy of Science* (trans. by A. Levin), University of Chicago Press, Chicago.

Huang, J.Y., Y.K. Wu, and H.Q. Ye. 1996. *Acta Mater*. 44:1211. Inoue, A. 1997. Private communication. Inturi, R.B., and Z. Szklavska-Smialowska. 1992. *Corrosion* 48:398. Karch, J., R. Birringer, and H. Gleiter. 1987. *Nature* 330:556.

Jena, P., S.N. Khanna, and B.K. Rao. 1996. In *Science and technology of atomically engineered materials*, ed. P. Jena. River Edge, NJ: World Scientific.

Jonas, H. The Imperative of Responsibility. In Search of an Ethics for the Technological Age. Chicago: University of Chicago Press, 1985.

Karak, N., and S. Maiti. 1997. Dendritic polymers: A class of novel material. *J. Polym. Mater*.14:105.

Karger, J., and D.M. Ruthven. 1992. *Diffusion in zeolites*. New York: J. Wiley. Krätschmer, W., L.D. Lamb, K. Fostiropoulos, and D.R. Huffman. 1990. *Nature* 347:354.

Karplus, M. & Weaver, D. L. (1976) *Nature (London)* 260, 404-406.

Kirschvink, J.L., A. Koyayashi-Kirschvink, and B.J. Woodford. 1992. Magnetite biomineralization in the human brain. *Proc. Nat'l. Acad. Sci.* USA 89:7683-7687.

Kishida, M., T. Fujita, K. Umakoshi, J. Ishiyama, H. Nagata, and K. Wakabayashi. 1995. *Chem. Commun*. 763.

Klein, J.D., et al. 1993. *Chem. Mater*. 5:902. Koch, C.C. 1989. Materials synthesis by mechanical alloying. *Annual Review of Mater. Sci*. 19:121-143.

Kourilsky, Ph.&Viney, G. Le Principe de précaution. Report to the Prime Minister, Paris, Éditions Odile Jacob, 2000.

Krejchi, M.T., E.D.T. Atkins, A.J. Waddon, M.J. Fournier, T.L. Mason, and D.A. Tirrell. 1994. Chemical sequence control of beta-sheet assembly in macromolecular crystals of periodic proteins. *Science* 265:1427-1432.

Krejchi, M.T., S.J. Cooper, Y. Deguchi, E.D.T. Atkins, M.J. Fournier, T.L. Mason, and D.A. Tirrell. 1997. Crystal structures of chain-folded antiparallel beta-sheet assemblies from sequence-designed periodic polypeptides. *Macromolecules* 30:5012-5024.

Kurzweil, R.: 1998, *The Age of Spiritual Machines, How We Will Live, Work and Think in the New Age of Intelligent Machines*, Phoenix, New York.

Kyprianidou-Leodidou, T., W. Caseri, and V. Suter. 1994. *J. Phys. Chem*. 98:8992. Kung, H.H., and E.I. Ko, 1996. *Chem. Eng. J*. 64:203.

Leonard, D., M. Krishnamurthy, C.M. Reaves, S.P. Denbaars, and P.M. Petroff. 1993. Direct formation of quantum-sized dots from uniform coherent islands of InGaAs on GaAs surfaces. *Appl. Phys. Lett*. 63:3203-5.

Liang, G., Z. Li, and E. Wang. 1996. *J. Mater. Sci*. Makino, A., A. Inoue, T. Hatanai, and T. Bitoh. 1997. *Materials Science Forum* 235-238: 723.

Mao, C., W. Sun, and N.C. Seeman. 1997. Construction of Borromean rings from DNA. *Nature* 386:137-138.

Martin, T.P., N. Malinowski, U. Zimmerman, U. Naher, and H. Schaber. 1993. *J. Chem. Phys*. 99:4210.

Martin, T.P., U. Naher, H. Schaber, U. Zimmerman. 1993. *Phys. Rev. Lett*. 70:3079. Prigogine, I., and S. Rice. 1988. *Advances in chemical physics*, Vol. 70, Parts 1 & 2. New York: J. Wiley.

McCammon, J. A., Gelin, B. R. & Karplus, M. (1977) *Nature (London)* 267, 585-590.

McConnell, H.M. 1996. Light-addressable potentiometric sensor: Applications to drug discovery. In *Nanofabrication and biosystems*, ed. Hoch et al.

McCulloch, W.S.: 1962, 'The imitation of one form of life by another – Biomimesis', in: E.E. Bernard & M.R. Kare (eds.), *Biological prototypes and synthetic systems*, Plenum Press, New York, vol. 1., p. 393-97.

Mehl, R.F., and R.W. Cahn. 1983. Historical development. In *Physical metallurgy*. North Holland. Milligan, W.W., S.A. Hackney, M. Ke, and E.C. Aifantis. 1993. *Nanostructured Materials* 2:267.

Minsky, M.: 1995, 'Virtual Molecular Reality', in: Krummenacker, M. & Lewis, J. (eds.), *Prospects in Nanotechology. Proceedings of the 1st general conference on nanotechnology: developments, applications, and opportunities, November 11-14, 1992, Palo-Alto*, John Wiley & Sons, New York, pp. 187-205.

Mishra, R.S., and A.K. Mukherjee. 1997. Oral presentation at TMS meeting, Indianapolis, Indiana, 16-18 September 1997, to be published in proceedings of Symp. "Mechanical Behavior of Bulk Nano-Materials."

Mishra, R.S., R.Z. Valiev, and A.K. Mukherjee. 1997. *Nanostructured Materials* 9:473. Morris, D.G., and M.A. Morris. 1991. *Acta Metall. Mater.* 39:1763-1779.

Moore, J.C., H.M. Jin, O. Kuchner, and F.H. Arnold. 1997. Strategies for the in vitro evolution of protein function: Enzyme evolution by random recombination of improved sequences. *J. Mol. Biol.* 272:336-347.

Morris, D.G., and M.A. Morris. 1997. *Materials Science Forum* 235-238:861. Nagpal, P., and I. Baker. 1990. *Scripta Metall. Mater.* 24:2381.

Nieman, G.W., J.R. Weertman, and R.W. Siegel. 1991a. *Mater. Res. Soc. Symp. Proc.* 206:581-586.

Nomura, M. & Held, W. (1974) in *Ribosomes,* eds. Nomura, M., Tissiers, A. & Lengyel, P. (Cold Spring Harbor Laboratory, Cold Spring Harbor, NY), pp. 193-203.

Nordmann, A. (rapp.) Converging Technologies—Shaping the Future of European Societies, European Commission report, 2004.

Parker, J.C., et al. 1995. U.S. Patent 5,460,701. POST (Parliamentary Office of Science and Technology). 1995.

Phoenix, C.; Drexler, E.: 2004, 'Safe Exponential Manufacturing', *Nanotechnology*, 15, 869-72.

Rademann, K., B. Kaiser, U. Even, F. Hensel. 1987. *Phys. Rev. Lett.* 59:2319. Rao, C.N.R., B.C. Satishkumar, and A. Govindaraj. 1997. *Chem. Commun.* 1581.

Ramanan, V.R. 1998. Nanocrystalline soft magnetic alloys for applications in electrical and electronic devices. In *R&D Status and Trends*, ed. Siegel et al.

Rao, M.B., and S. Sircar. 1993. *Gas Separation and Purification* 7:279. Reetz, M.T., et al. 1995. *Science* 267:367.

Rofagha, R., R. Langer, A.M. El-Sherik, U. Erb, G. Palumbo, and K.T. Aust. 1991. *Scr. Metall. Mater.* 25:2867.

Roher, H. 1993. *Jpn. J. Appl. Phys.* 32:1335. Rohlfing, E.A., D.M. Cox, and A. Kaldor. 1984. *J. Chem. Phys.* 81:3846.

Romanov, A.E., V.I. Vladimirov. 1992. In *Dislocations in solids*, ed. F.R.N. Nabarro, Vol. 9. Amsterdam: North-Holland. Salishekev, G.A., O.R. Valiakhmetov, V.A. Valitov, and S.K. Mukhtarov. 1994. *Materials Science Forum*. 170-172:121.

Ruthven, D.M., S. Farooq, K.S. Knaebel. 1994. *Pressure swing adsorption*. New York: VCH Publishers. Sayari, A. 1996. *Chem. Mater.* 8:1840.

S. P. Ho and W. F. DeGrado, "Design of a 4-Helix bundle protein: Synthesis of peptides which self-associate into a helical protein." *J. Am. Chem. Soc.,* 109: 6751-6758 (1987).

Sanders, P.G., J.A. Eastman, and J.R. Weertman. 1996. In *Processing and properties of nanocrystalline materials*, ed. Suryanarayana et al.

Sanders, P.G., M. Rittner, E. Kiedaisch, J.R. Weertman, H. Kung, and Y.C. Lu. 1997. *Nanostructured Mater.* 9:433.

Sankey, H.: 1994, *The Incommensurability Thesis*, Avebury, Aldershot.

Sarikaya, M.; Aksay, I. (eds.), 1995, *Biomimetics. Design and Processing of Materials*, AIP Press, Woodbury, New York.

Scanlan, R.M., W.A. Fietz, and E.F. Koch. 1975. *J. Appl. Phys.* 46:2244.

Schiefsky, M.J.: forthcoming, 'Art and Nature in Ancient Mechanics', in: W.R. Newman & B.

Bensaude-Vincent (eds.), *The Artificial and the Natural: An Ancient Debate and its Modern Descendants*, MIT Press, Cambridge, MA.

Schultz, L., J. Wecker, and E. Hellstern. 1987. *J. Appl. Phys.* 61:3583.

Schultz, L., K. Schnitzke, and J. Wecker. 1989. *J. Magn. Mater.* 80:115.

Schwarz, R.B. 1998. Storage of hydrogen powders with nanosized crystalline domains. In *R&D Status and Trends*, ed. Siegel et al.

Shen, T.D., C.C. Koch, T.Y. Tsui, and G.M. Pharr. 1995. *J. Mater. Res.* 10: 2892.

Shull, R.D., R.D. McMichael, and J.J. Ritter. 1993. *Nanostructured Mater.* 2:205.

Siegel et al. Ogunnaike, B and W. Ray. 1994. *Process dynamics, modeling and control.* Oxford University Press, pp 5-21; 1033-48.

Siegel R.W. and G.E. Fougere. 1994. In *Nanophase materials*, ed. G.C. Hadjipanayis and R.W. Siegel. Netherlands: Kluwer Acad. Publ.

Siegel, H.: 1980, 'Objectivity, Rationality, Incommensurability, and More', *British Journal for the Philosophy of Science*, 31, 359-384.

Siegel, R.W., E. Hu, and M.C. Roco, eds. 1998. *R&D status and trends in nanoparticles, nanostructured materials, and nanodevices in the United States.* Baltimore: Loyola College, International Technology Research Institute.

Smalley, R.: 2001, 'Of Chemistry, Love and Nanobots', *Scientific American*, 285, 76-77.

Smalley, R.: 2003b, 'Smalley Concludes', *Chemical & Engineering News,* 81, 41-42.

Smalley, R.E.: 2001, 'Of Chemistry, Love and Nanobots', *Scientific American*, (Sept.), 76-77.

3

Microsocpy, Nanoengineering & Nanomanipulation: An Overview

SPM and Nanomanipulation

The Scanning Tunelling Microscope (STM) was invented by Binnig and Rohrer at the IBM Zürich laboratory in the early 1980s, and won them a Nobel Prize four years later. The principles of the instrument can be summarized with the help. A sharp conducting probe, typically made from tungsten is placed very close to a sample, that must also be a conductor. The tip is biased with respect to the sample as shown in the figure. The tip can be moved towards or away from the sample, i.e., in the $-z$ or $+z$ directions in the figure, by means of piezoelectric actuators. At Ångstrom-scale distances, a quantum-mechanical effect, called *tunneling*, causes electrons to flow across the tip/ sample gap, and a current can be detected. To first approximation, the tunneling current depends exponentially on the distance between tip and sample. This current is kept constant by a feedback circuit that controls the piezoelectric actuators. Because of the current/distance relationship, the distance is also kept constant, and the z accuracy is very high because any small z variation causes an exponential error in the current. Additional piezo motors drive the tip in a xy scanning motion. Since the tip/ sample gap is kept constant by the feedback, the scanning tip traverses a surface parallel to the sample surface. The result of a scan is a $z(x, y)$ terrain map, with enough resolution to detect atomic-scale features of the sample, as indicated diagrammatically in the figure. Various instruments analogous to the STM have been built. They exploit physical properties other than the tunneling effect on which the STM is based. The most common of these other instruments is the Atomic Force Microscope (AFM), which is based on interatomic forces. All of these instruments are collectively known as Scanning Probe Microscopes (SPMs). The principles of operation of the AFM.. The forces between atoms in the tip and sample cause a deflection of the cantilever that carries the tip. The amount of deflection is measured by means of a laser beam bouncing off the top of the cantilever. (There are other schemes for measuring deflection.) The force depends on the tip/sample gap, and therefore servoing on the force ensures that the distance is kept constant while scanning, as in the STM. The AFM does not require conducting tips and samples, and therefore has wider applicability than the STM. If the tip of the AFM is brought very close to the sample, at distances of a few Å, repulsive forces prevent the tip from penetrating the sample. This mode of operation is called contact mode. It provides good resolution but cannot be used with delicate samples, *e.g.* biomaterials, which are damaged by the contact forces. Alternatively, the AFM tip can be placed at distances in the order of several nm or tens of nm, where the interatomic forces between tip and sample are attractive. The tip is vibrated at a frequency near the resonance frequency of the cantilever, in the kHz range. The tip/sample force is equivalent to a change in the spring constant of the cantilever, and causes a change in its resonance frequency. This change can be used as the error signal in the feedback circuit that controls the tip. (There are alternative detection schemes.) This mode of operation is called non-contact mode. It has poorer resolution than contact mode but can be used with delicate samples. Although the SPM is just about twenty years old, it has had a large scientific impact.

Since the early days of the SPM it was known that tip/sample interaction could produce changes in both tip and sample. Often these were undesirable, for example, a blunt probe due to a crash into the sample. But it soon became clear that one could produce new and desirable features on a sample by using the tip in a suitable manner. One of the first demonstrations was done by Becker and co-workers at Bell Labs, who managed to create nanometer-scale germanium structures on a germanium surface by raising the voltage bias of an STM tip. Much of the subsequent work falls under the category of nanolithography and will not be discussed here. In the following subsections we survey nanomanipulation research involving the SPM.

Pushing and pulling operations are not widely used in macrorobotics, although there has been interesting work on orienting parts by pushing, done by Matt Mason at CMU, Ken Goldberg at USC, and others. The techniques seem suitable for constructing 2-D structures. Interatomic attractive forces were used by Eigler *et al.* at IBM Almadén to precisely position xenon atoms on nickel, iron atoms on copper, platinum atoms on platinum, and carbon monoxide molecules on platinum. The atoms are moved much like one displaces a small metalic object on a table by moving a magnet under the table. The STM tip is placed sufficiently close to an atom for the attractive force to be larger than the resistance to lateral movement. The atom is then pulled along the trajectory of the tip. Eigler's experiments were done in ultra high vacuum (UHV) at very low temperature (4K). Low temperature seems essential for stable operation. Thermal noise destroys the generated patterns at higher temperatures. Lateral repulsive forces were used by Güntherodt's group at the University of Basel to push fullerene (C_{60}) islands of ~ 50nm size on flat terraces of a sodium chloride surface, in UHV, at room temperature, with a modified AFM. The ability to move the islands depends strongly on species/substrate interaction, e.g., C_{60} does not move on gold, and motion on graphite destroys the islands. Lateral forces opposing the motion are analogous to friction in the macroworld, but cannot be modeled simply by Coulomb friction or similar approaches that are used in macrorobotics. Mo at IBM Yorktown rotated pairs of antimonium atoms between two stable orientations 90 degrees apart. This was done in UHV at room temperature, on a silicon substrate, by scanning with an STM tip with a higher voltage than required for imaging. The rotation was reversible, although several scans were sometimes necessary to induce the desired motion. Samuelson's group at the University of Lund succeeded in pushing galium arsenide (GaAs) nanoparticles of sizes in the order of 30 nm on a GaAs substrate at room temperature in air. The sample is first imaged in non-contact AFM mode. Then the tip is brought close to a nanoparticle, the feedback is turned off and the tip is moved against the nanoparticle. Schaefer *et al.* at Purdue University push gold clusters with an AFM in a nitrogen environment at room temperature. They first image the clusters in non-contact mode, then remove the tip oscillation voltage, and sweep the tip across the particle in contact with the surface and with the feedback disabled.

A similar technique is being used at USC's Laboratory for molecular robotics to push colloidal gold nanoparticles with 15 nm diameters on a mica substrate at room temperature and in ambient air. We image in non-contact mode, then disable the feedback and push by moving the tip in a single line scan, without removing the tip oscillation voltage. The right a "USC" pattern written with gold nanoparticles. The z coordinate is encoded as brightness in this figure. On the left is the original random pattern, before manipulation with the AFM. Smaller objects have been arranged into prescribed patterns at room temperature by Gimzewski's group at IBM's Zürich laboratory. They push molecules at room temperature in UHV by using an STM. They have succeeded in pushing porphyrin molecules on copper, and more recently they have arranged bucky balls (i.e., C_{60}) in a linear pattern, using an atomic step in the copper substrate as a guide. C_{60} molecules on silicon also have been pushed with an STM in UHV at room temperature by Maruno *et al.* in Japan, and Beton *et al.* in the U.K. In Maruno's approach the STM tip is brought closer to the surface than in normal imaging mode, and then scan across a rectangular region with the feedback turned off. This causes many probe crashes. In Beton's approach the tip also is brought close to the surface, but the sweep is done with the feedback on and a high value for the tunneling current. Their success rate is in the order of only 1 in 10 trials.

Much of industrial macrorobotics is concerned with pick and place operations, which typically do not require very precise positioning or fine control. There are a few examples of experiments in which atoms or molecules are transferred to SPM tips, these are moved, and the atoms transferred back to the surfaces. Eigler *et al.* succeeded in transferring xenon atoms from platinum or nickel surfaces to an STM tip by moving the tip sufficiently close for the adsorption barriers of surface and tip to be comparable. An atom may leave the surface and become adsorbed to the tip, or vice-versa. Benzene molecules also have been transferred to and from tips. Eigler's group also has been able to transfer xenon atoms between an STM tip and a nickel surface by applying voltage pulses to the tip. This is attributed to electromigration, caused by the electric current flowing in the tunneling junction. All of Eigler's work has been done in UHV at 4K. Avouris' group, at IBM Yorktown, and the Aono group in Japan have transferred silicon atoms between a tungsten tip and a silicon surface in UHV at room temperature, by applying voltage pulses to the tip. The mechanism for the transfer is believed to be field-induced evaporation, perhaps aided by chemical phenomena at the tip/surface interface in Avouris work.

This is the most sophisticated form of macrorobotic motion. It involves fine motions, as in a peg-in-hole assembly, in which there is accomodation and often force control, for example to ensure contact between two surfaces as they slide past each other. Compliance is crucial for successful assembly operations in the presence of spatial and other uncertainties, which are unavoidable. The study of the nanoscale analog of compliant motion seems to be virgin territory. We speculate that the analog of compliance is chemical affinity between atoms and molecules. It is suspected that such "chemical compliance" may prove essential for nanoassembly operations at room temperature, in the presence of thermal noise. It seems likely that successful assembly of nanoscale components will require a combination of precise positioning and chemical compliance. Therefore, work on self-assembling structures is relevant.

To first approximation, an SPM is a 3 degree-of-freedom robot. It can move in x, y, and z, but cannot orient its tip, which is the analog of a macrorobotic hand. (How could a 6 degree-of-freedom SPM be built, and what could be done with it, are interesting issues.) The vertical displacement is controlled very accurately by a feedback loop involving tunneling current or force (or other quantities for less-common SPMs). But nanoscale x, y motion over small regions (e.g., within a 5 micron square) is primarily open-loop, because of a lack of suitable sensors that can be used in a feedback scheme. Accurate horizontal motion relies on calibration of the piezoelectric actuators, which are known to suffer from a variety of problems such as creep and hysteresis. In addition, thermal drift of the instrument is very significant. At room temperature a drift of one atomic diameter per second is common, which means that manipulation of atomic objects on a surface is not unlike picking parts from a conveyor belt. Thermal drift is negligible if the SPM is operated at very low temperatures, and all the experiments in atomic-precision manipulation to date have been done at 4K. This involves complex technology and is clearly undesirable. For room-temperature nanomanipulation, drift, creep and hysteresis must be taken into account. Ideally, compensation should be automatic. Research is under way at USC on how to move accurately an SPM tip in the presence of all these sources of error. To complicate matters further, the SPM often must operate in a liquid environment, especially if the objects to be manipulated are biological. Little is known about nanomanipulation in liquids.

The SPM functions both as a manipulator and a sensing device. The lack of a direct and independent means of establishing "ground truth" while navigating the tip causes delicate problems. The SPM commands are issued in instrument or robot coordinates, and sensory data also are acquired in robot coordinates, whereas manipulation tasks are expressed in sample or task coordinates. These two coordinate systems do not coincide, and indeed are in relative motion due to drift. Accurate motion in task coordinates may be achieved by a form of visual servoing, by tracking features of the sample and moving relative to them. We move the tip to the approximate position of a nanoparticle to be pushed, and then search for it through single line scans. Feature tracking assumes that features are stationary in task coordinates. This assumption may fail at room temperature if the features are sufficiently small, because of thermal agitation. Hence, atomic manipulation at room temperature may require artificially-introduced features that can be tracked to establish a task coordinate system. It must also deal with the spatial uncertainty associated with the thermal motion of the atoms to be moved. Larger objects such as molecules and clusters have lower spatial uncertainty, and should be easier to handle.

The SPM output signal depends not only on the topography of the sample but also on the shape of the tip, and on other characteristics of the tip/sample interaction. For example, the tunneling current in an STM depends on the electronic wavefunctions of sample and tip. To first approximation one may assume that the tip and sample are in contact. Under this assumption one can use configuration-space techniques to study the motion of the tip]. The procedure in 2-D. On the top of the figure we consider a tip with a triangular end and a square protrusion in the sample. In the bottom we consider a tip with a semi-circular end and the same protrusion. We choose as reference point for the tip its apex, and reflect the tip about the reference point. The configuration-space obstacle that corresponds to the real-space protrusion obstacle is obtained by sweeping the inverted tip over the protrusion so that the apex remains inside the obstacle. Mathematically, we are calculating the Minkowski sum of the inverted tip and the protrusion. The path of the tip in its motion in contact with the obstacle is the detected topographical signal. As shown in the figure, the sensed topography has been broadened by the dimensions of the tip. Note, however, that the detected height is correct. (Minkowski operations and related mathematical morphology tools were introduced in the SPM literature only recently.) Tip effects are sometimes called "convolution", by analogy with the broadening of an impulse passing through a linear system, and one talks of "deconvolving" the image to remove tip effects. The configuration-space analysis outlined above is purely geometric and provides only a coarse approximation to the SPM output signal. More precise calculations may be performed numerically. For example, in contact AFM we can assume specific atomic distributions for the tip and sample, and a specific form for the interatomic forces, and compute the resulting tip/sample force. A major issue in tip-effect compensation is that the shape of the probe is not known, and indeed may vary during operation. For example, atoms may be adsorbed on the tip or lost because of the contact with the sample. The most promising approach for dealing with tip effects consists of estimating the tip shape by using it to image known features. If necessary, artificial features may be introduced into the scene for tip estimation purposes. The estimated tip shape can then be used to remove (at least in part) the tip effects from the image by using Minkowski operations. Removal procedures that take into account more sophisticated, non-geometric effects do not appear to be known.

Sensor fusion techniques may be used, at least in principle, for increasing the quality of the sensory data, because it is possible to access several signals during an SPM scan. For example, vertical and lateral force (akin to friction) can be recorded simultaneously in typical AFMs. To our knowledge, sensor fusion has not been attempted in SPM technology. It is clear that faithful sensory data should facilitate manipulation tasks. What is not clear is whether clever manipulation strategies can compensate for the imperfections of SPM data.

The SPM tip is the primary end effector in nanomanipulation. A plain, sharp tip seems to be adequate for most pushing operations. In some cases it may also suffice for picking and depositing objects, especially in conjunction with electrostatic forces generated by applying a suitable bias to the tip. This requires both a conducting tip and a conducting substrate. Pick-and-place tasks, however, usually require the nanoscale analog of a gripper. Very little is known about molecular grippers. One can think of a nanogripper as a molecule or cluster of molecules that are attached to a tip and are capable of picking up and transporting other molecules or particles. (Tips with attached molecules are said to be functionalized.) Ideally these grippers should be switchable, so as to pick and release objects on command. Candidates for grippers are certain molecules such as cyclodextrins, which have large cavities that can carry other molecules. Coating a tip with strands of DNA may also permit picking up objects that are coated with complementary strands. In both of these examples, switching the gripper on and off is not a solved problem. Techniques for changing SPM tips automatically do not exist. Changes must be done manually, and it is very difficult or impossible to maintain sample registration, *i.e.*, to return to the same position on the sample after tip replacement. This implies that one often must image using a tip with an attached gripper. Again, little is known about imaging with such functionalized tips.

In the macroworld fixtures are often necessary to hold mechanical parts during assembly and other manufacturing operations. In micromechanics sacrificial layers are used as scaffolding to fabricate certain microelectromechanical systems (MEMS). At the nanoscale, the analogs of fixtures are substrates that ensure that certain objects remain fixed during manipulation, while others are allowed to move. Substrate selection seems to be highly dependent on the chemistry of the objects being manipulated.

In macrorobotics the physical processes involved are mechanical and relatively well understood. At the nanoscale, the processes are chemical and physical, and still an area of active research. In addition, nanomanipulation takes place in several different environments, such as liquids, air or UHV. The environment has a strong influence on the physics and chemistry of the processes. Nanomanipulation is not restricted to mechanical interactions. For example, light, electrostatic fields, and the pH of a liquid all are candidates for controlled interaction with nanoparticles.

High level programing and planning systems are highly desirable, and indeed essential for assembling complex structures. One must begin with relatively low-level programing primitives and build upon them constructs at a higher level of abstraction. High-level commands must be compiled into low-level primitives. This compilation may involve sophisticated computations, for example to ensure collision-free paths in an environment with large spatial uncertainty. What are the relevant high-level manipulation tasks? For example, what is the nanoscale equivalent of a peg-in-hole insertion? In short, we may need to adapt much of what is known about macrorobotics to the nanoworld. It is likely that new concepts will also be needed, because the physics and chemistry of the phenomena and objects are quite different in these two worlds. What hardware primitives are suitable as building blocks? The nanotechnology literature suggests hardware primitives based on DNA structures such as those built in Seeman's lab, proteins, and diamondoid structures. Biomaterials such as DNA and proteins may be too flimsy, whereas diamondoid structures are expected to be very strong. No experiments have yet been reported in which any of these components are successfully assembled into a composite structure. Which tasks should one attempt first? What should one try to build? Here the options are in the realms of electronics, photonics, mechanics, or biomaterials. On-going research at USC is attempting to build nanowires as assemblies of gold particles, and arrays of nanostructures with photonic properties on patterned semiconductor substrates.

Nanorobotics manipulation with SPMs is a promising field that can lead to revolutionary new science and technology. But it is clearly in its infancy. Typical nanomanipulation experiments reported in the literature involve a team of very skilled, Ph.D.-level researchers working for many hours in a tightly-controlled environment (typically in ultra high vacuum and at low temperature, often 4K) to build a pattern with tens of nanoparticles. It still takes the best groups in the world some 10 hours to assemble a structure with about 50 atoms. This is simply too long-changes will occur in many systems, *e.g.* contamination or oxidation of the components, on a timescale that will constrain the maximum time available for nanomanipulation. Requiring all operations to take place at 4K and in UHV also is not practical for widespread use of nanomanipulation. In short, nanomanipulation today is more of an experimental tour-de-force than a technique that can be routinely used. It is clear that complex tasks cannot be accomplished unless the SPM is commanded at a higher level of abstraction. Compensation for instrument inaccuracies should be automatic, and the user should be relieved from many low-level details. Building a high level programing system for nanomanipulation is a daunting task. The various component technologies needed for nanomanipulation must be developed and integrated. These technologies include: substrates that serve as nanofixtures or nanoworkbenches on which to place the objects to be manipulated; tips, probes and molecules that serve as grippers or end-effectors; chemical and physical nanoassembly processes; primitive nanoassembly operations that play a role analogous to macroassembly tasks such as peg-in-hole insertion; methods for exploiting self-assembly to combat spatial uncertainty, in a role analogous to mechanical compliance in the macroworld; suitable hardware primitives for building nanostructures; and algorithms and software for sensory interpretation, motion planning, and driving the SPM. This is a tall order, and requires an interdisciplinary approach that combines synergistically the knowledge and talents of roboticists and computer scientists with those of physicists, chemists, materials scientists, and perhaps biologists. SPM-based assembly methods face a major scale-up challenge. Building complex structures one atom (or even one nanoparticle) at a time is very time consuming. We believe that SPMs will have applications in the exploration of new structures, which may later be mass produced by other means. This is the nanoworld analog of rapid protyping technologies such as stereolithography that are becoming popular at the macroscale.

There are at least two approaches for fighting the serial nature of SPM manipulation. The first involves the use of large arrays of SPMs on a chip. These chips are being developed at Cornell and Stanford. Programing such arrays for coordinated assembly tasks poses interesting problems. The second approach is subtler, and consists of using the SPM to construct structures that are capable of self-replication. The best known such structures involve DNA, but other systems also exist. Self-replication is inherently an exponential process. In summary, nanomanipulation with SPMs may have a revolutionary impact on science, technology, and the way we live. To fully exploit its potential we will have to develop powerful systems for programing nanorobotic tasks. Much of what is known in macrorobotics is likely to be relevant, but may have to be adapted to the nanoworld, where phenomena and structures are quite different from their macroscopic counterparts. Research at USC and elsewhere is progressing, with promising results. Nanomanipulation, perhaps coupled with self-assembly, is expected to eventually succeed in building true nanorobots, *i.e.*, devices with overal dimensions in the nanometer range and capable of sensing, "thinking", and acting. Complex tasks are likely to require a group of nanorobots working cooperatively. This raises interesting issues of control, communications, and programing of robot "societies".

Scanning Probe Microscopy Techniques (SPMT)

Scanning Probe Microscopy has enabled researchers to image surfaces at the nanometer scale. Rather than using a beam of light or electrons, SPM uses a fine probe that is scanned over a surface (or the surface is scanned under the probe). By using such a probe, researchers are no longer restrained by the wavelength of light or electrons. The resolution obtainable with this technique can resolve atoms, and true 3-D maps of surfaces are possible. Scanning Probe Microscopy is a general term, used to describe a growing number of techniques that use a sharp probe to scan over a surface and measure some property of that surface. Some examples are STM (scanning tunneling microscopy), AFM (atomic force microscopy), and NSOM (Near-Field Scanning Optical Microscopy).

Scanning Tunneling Microscopy (STM)

The development of the family of scanning probe microscopes starts with the original invention of the STM in 1981. Gerd Binnig and Heinrich Rohrer developed the first working STM while working at IBM Zurich Research Laboratories in Switzerland. This instrument would later win Binnig and Rohrer the Nobel prize in physics in 1986. The STM works by scanning a very sharp metal wire tip over a surface. By bringing the tip very close to the surface, and by applying an electrical voltage to the tip or sample, we can image the surface at an extremely small scale – down to resolving individual atoms. The STM is based on several principles. One is the quantum mechanical effect of tunneling. It is this effect that allows us to "see" the surface. Another principle is the piezoelectric effect. It is this effect that allows us to precisely scan the tip with angstrom-level control. Lastly, a feedback loop is required, which monitors the tunneling current and coordinates the current and the positioning of the tip.

Tunneling is a quantum mechanical effect. A tunneling current occurs when electrons move through a barrier that they classically shouldn't be able to move though. In classical terms, if you don't have enough energy to move "over" a barrier, you won't. However, in the quantum mechanical world, electrons have wavelike properties. These waves don't end abruptly at a wall or barrier, but taper off quite quickly. If the barrier is thin enough, the probability function may extend into the next region, though the barrier! Because of the small probability of an electron being on the other side of the barrier, given enough electrons, some will indeed move through and appear on the other side. When an electron moves though the barrier in this fashion, it is called tunneling. Quantum mechanics tells us that electrons have both wave and particle like properties. Tunneling is an effect of the wavelike nature. The top image shows us that when an electron (the wave) hits a barrier, the wave doesn't abruptly end, but tapers off very quickly—exponentially. For a thick barrier, the wave doesn't get past. The bottom image shows the senario if the barrier is quite thin (about a nanometer). Part of the wave does get through, and therefore some electrons may appear on the other side of the barrier.. Because of the sharp decay of the probability function through the barrier, the number of electrons that will actually do this is very dependent upon the thickness of the barrier. The actual current through the barrier drops off exponentially with the barrier thickness. To extend this description to the STM: The starting point of the electron is either the tip or sample (depending on the setup of the instrument). The barrier is the gap (air, vacuum, liquid), and the second region is the "other side" – tip or sample, again, depending on the experimental setup. By monitoring the current through the gap, we have very good control of the tip-sample distance.

The piezoelectric effect was discovered by Pierre Curie in 1880. The effect is created by squeezing the sides of certain crystals, such as quartz or barium titanate. The result is the creation of opposite charges on the sides. The effect can be reversed as well; by applying a voltage across a piezoelectric crystal, it will elongate or compress. These materials are used to scan the tip in an STM, and most other scanning probe techniques. A typical piezoelectric material used in STMs is PZT (Lead Zirconium Titanate).

Obviously, you need electronics to measure the current, scan the tip, and translate this information into a form that we can use. A feedback loop constantly monitors the tunneling current and makes adjustments to the tip to maintain a constant tunneling current. These adjustments are recorded by the computer and presented as an image in the STM software. Such an setup is called a "constant current" image. In addition, for very flat surfaces, the feedback loop can be turned off and only the current is displayed. This is a "constant height" image.

The STM is cabable of acquiring remarkable images on the most extreme scale, easily resolving atomic structure in the right environments. For some very interesting work with STM, see the IBM image gallery.

This STM image shows the direct of standing-wave patterns in the local density of states of the Cu(111) surface. These spatial oscillations are quantum-mechanical interference patterns caused by scattering of the two-dimensional electron gas off the Fe adatoms and point defects.

Atomic Force Microscopy (AFM)

The Atomic Force Microscope was developed to overcome a basic drawback with STM—that it can only image conducting or semiconducting surfaces. The AFM, however, has the advantage of imaging almost any type of surface, including polymers, ceramics, composites, glass, and biological samples. Binnig, Quate, and Gerber invented the Atomic Force MIcroscope in 1985. Their original AFM consisted of a diamond shard attached to a strip of gold foil. The diamond tip contacted the surface directly, with the interatomic van der Waals forces providing the interaction mechanism. Detection of the cantilever's vertical movement was done with a second tip—an STM placed above the cantilever.

Today, most AFMs use a laser beam deflection system, introduced by Meyer and Amer, where a laser is reflected from the back of the reflective AFM lever and onto a position-sensitive detector. AFM tips and cantilevers are microfabricated from Si or Si_3N_4. Typical tip radius is from a few to 10s of nm.

Measuring forces

Because the AFM relies on the forces between the tip and sample, knowing these forces is important for proper imaging. The force is not measured directly, but calculated by measuring the deflection of the lever, and knowing the stiffness of the cantilever. Hook's law gives F = -kz, where F is the force, k is the stiffness of the lever, and z is the distance the lever is bent.

Because of AFM's versatility, it has been applied to a large number of research topics. The Atomic Force Microscope has also gone through many modifications for specific application requirements.

The first and foremost mode of operation, contact mode is widely used. As the tip is raster-scanned across the surface, it is deflected as it moves over the surface corrugation. In constant force mode, the tip is constantly adjusted to maintain a constant deflection, and therefore constant height above the surface. It is this adjustment that is displayed as data. However, the ability to track the surface in this manner is limited by the feedback circuit. Sometimes the tip is allowed to scan without this adjustment, and one measures only the deflection. This is useful for small, high-speed atomic resolution scans, and is known as variable-deflection mode.

Because the tip is in hard contact with the surface, the stiffness of the lever needs to be less that the effective spring constant holding atoms together, which is on the order of 1-10 nN/nm. Most contact mode levers have a spring constant of < 1N/m.

Lateral Force Microscopy

LFM measures frictional forces on a surface. By measuring the "twist" of the cantilever, rather than merely its deflection, one can qualitatively determine areas of higher and lower friction.

Noncontact mode belongs to a family of AC modes, which refers to the use of an oscillating cantilever. A stiff cantilever is oscillated in the attractive regime, meaning that the tip is quite close to the sample, but not touching it (hence, "noncontact"). The forces between the tip and sample are quite low, on the order of pN (10^{-12} N). The detection scheme is based on measuring changes to the resonant frequency or amplitude of the cantilever.

Commonly referred to as "tapping mode" it is also referred to as intermittent-contact or the more general term Dynamic Force Mode (DFM).

A stiff cantilever is oscillated closer to the sample than in noncontact mode. Part of the oscillation extends into the repulsive regime, so the tip intermittently touches or "taps" the surface. Very stiff cantilevers are typically used, as tips can get "stuck" in the water contamination layer. The advantage of tapping the surface is improved lateral resolution on soft samples. Lateral forces such as drag, common in contact mode, are virtually eliminated. For poorly adsorbed specimens on a substrate surface the advantage is clearly seen.

Force modulation refers to a method used to probe properties of materials through sample/tip interactions. The tip (or sample) is oscillated at a high frequency and pushed into the repulsive regime. The slope of the force-distance curve is measured which is correlated to the sample's elasticity. The data can be acquired along with topography, which allows comparison of both height and material properties.

In Phase mode imaging, the phase shift of the oscillating cantilever relative to the driving signal is measured. This phase shift can be correlated with specific material properties that effect the tip/sample interaction. The phase shift can be used to differentiate areas on a sample with such differing properties as friction, adhesion, and viscoelasticity. The techniques is used simultaneously with DFM mode, so topography can be measured as well.

SNOM offers the use of a very small light source as the imaging mechanism. By using a quasipoint light source with a diameter much smaller than the wavelength of light, one can achieve resolutions better than the diffraction limit. The probe, however, must be very close to the surface; much closer than the wavelength of the light. This region is the "Near-Field" and hence the name of the technique. Typically, laser light is fed to the aperture via an optical fiber. The aperture can be a tapered fiber coated with a metal (such as Al), a microfabricated hollow AFM probe, or a tapered pipette. Normally, the size of the point light source determines the resolution obtainable. There are two types of feedback typically used to maintain the proper working distance of the probe to the sample. One method is quite similar to how feedback works with an AFM—by using a cantilevered probe, the normal force is monitored, typically by using a beam-deflection setup as in most AFMs. The second method uses a tuning fork. By attaching the fiber to a tuning fork, which oscillates at its resonant frequency, can can monitor changes in the amplitude as the tip moves over the surface. The tip is moved laterally, and this techniques is normally referred to as "shear-force" feedback. Depending upon the sample being imaged, there are multiple modes of operation for NSOM.

Transmission: Lightsource travels through the probe aperture, and transmits through sample. Requires a transparent sample.

Reflection: Lightsource travels through the probe aperture, and reflects from the surface. Lower light intensity, and tip-dependent, but allows for opaque samples.

Collection: Sample is illuminated from large outside light source, and the probe collects the reflected light.

Illumination/Collection: The probe both illuminates the sample and collects the reflected light. Detection of the signal can be handled a number of different ways:

Spectrometer, APD (Avalanche Photo Diode), Photomultiplier Tube, or CCD

Data Display: several contrast mechanisms in NSOM can be used, including polarization, topography, birefringence, index of refraction, fluorescence, wavelength dependence, and reflectivity.

An example of a commerical NSOM is the WITec AlphaSNOM.

The Nanosurf® EasyScan 2 AFM and STM

Configure your own AFM or STM that suits your needs at an affordable cost. The highly modular easyScan 2 line allows you to move from a very affordable STM system to a highly functional multiple mode AFM/STM package. You can configure what you need, or you can add options later. The Swiss-made system is manufactured near the birthplace of the STM. High quality, precision machining, and a smart easy-to-use design makes the easyScan 2 the perfect choice for teaching or research. The easyScan 2 system is built around the elegant easyScan 2 controller. The system and software can be configured to control either of 2 STM microscope heads, or any of 3 different AFM heads. Auxiliary boards and functional upgrades are available to expand the system's capabilities.

The easyScan 2 STM:

The simple elegance and affordability of the STM has been experienced by hundreds of researchers and educators around the world. The easyScan 2 offers new capabilities not available with the first model, including many upgrade options.

The easyScan 2 AFM:

One can easily upgrade the STM to the AFM, or simply bypass the STM option altogether. The new easyScan 2 AFM lets you get started with atomic force microscopy at a low cost, and you have the ability to upgrade functions later one.

Advanced easyScan 2 options:

The easyScan 2 AFM can be upgraded by adding dynamic force functions, dual-view video, sample translation, a signal access monitor, software capabilities and many other options.

4

Nano-electromechanical Systems: Applications & Importance

Nonopolis Indusrial Partners

SPECTRO will launch its next generation, high performance CIROS VISION ICP-OES CCD optical emission spectrometer at Pittcon 2004, 8-11 March in Chicago, USA. The new SPECTRO CIROS VISION ICP system is the most powerful, flexible and fastest ICP spectrometer in the world today and is the only system with a broad emission range to enable complete chemical elemental analysis. New highlight features and functionality include a significantly faster readout system, an improved Plasma Interface, improved capabilities for unattended control, and a brand new software platform. The new generation SPECTRO CIROS VISION ICP delivers a comprehensive range of new benefits to users including:

- new readout system—increases data processing speeds by a factor of ten, enabling real time standardization without loss of dynamic range, further improving precision and stability for the analysis of higher concentrations
- enhanced Plasma Interface—easier to handle and maintain while giving higher linearity
- optimized fluid paths—yielding shorter analysis times
- high sample throughput and 24 hour continuous operation
- extended control mechanisms, email alert function, and installation of WebCams for observation of the plasma offer excellent monitoring and diagnosis tools for unattended operation
- SMART ANALYZER VISION software platform – compliant to US FDA 21 CFR Part 11, the new 32-bit application offers unique reprocessing capabilities that enable elements to be added, method parameters to be edited and changed, even methods created and previously measured samples analyzed.

SPECTRO—the global leader in Optical Emission and XRF Spectroscopy—is a multinational company committed to innovation, instrumentation support and service. With more than 400 employees, SPECTRO has shipped over 20,000 instruments to customers throughout the world. SPECTRO is committed to producing advanced products, better and more diverse solutions, and providing an unparalleled support infrastructure. For more information visit www.spectro.com.

Nano & Micro Electromechanical Systems

Scientific innovator in high-resolution ultrasonic spectroscopy (HR-US) for material analysis will be unveiling its ground-breaking NEW high-resolution ultrasonic instrument, the Titration Kinetic Analyser on booth A4.220, Hall A4 at Analytica 2004, 11-14 May 2004 in Munchen, Germany. Ultrasonic Scientific is also showcasing its new customer-driven spectrometry software, HR-US accessories, and is holding on-booth seminars on how High-Resolution Ultrasonic Spectroscopy can be used for solving some of the main analytical problems facing the pharmaceutical, food, and personal care and polymer industries.

Ultrasonic Scientific's new Titration Kinetic Analyser will be launched in Europe at Analytica 2004. This new high-resolution ultrasonic instrument measures microstructural and molecular processes in liquids when concentration of selected compounds is changed (titration), or with time, when a chemical reaction is started by an injection of a required compound.

High-resolution ultrasonic spectroscopy is a novel technique for material analysis employing high-frequency acoustical (ultrasonic) waves. The application of ultrasound is well known; in medicine, ultrasound is used to visualize the internal organs of a patient's body. It is also used by submarines for underwater navigation.

Nanomaterials

Patented Process for Design, Development, and Production of Titanium Oxide Structures Customized for Titanium Metal Production

Altair Nanotechnologies, Inc. a developer and manufacturer of innovative nanomaterial products, was awarded its fourth patent entitled "Method for Producing Catalyst Structures" by the U.S. Patent Office. Patent number 6689716 was issued on February 10, 2004. The inventors of the process are Dr. Bruce J. Sabacky and Timothy M. Spitler, both of Altair Nanotechnologies.

"This patent, which further differentiates us within the rapidly growing nanotechnology market, underscores our commitment to bring nanotechnology out of the laboratory and into commercialization," said Altair President Dr. Rudi E. Moerck. Titanium is the 10th most abundant element in the earth's crust and the fourth most abundant structural metal. Titanium metal has unique physical and chemical properties including high strength-to-weight ratio, good corrosion resistance and excellent ballistic armor protection while offering significant weight reduction over conventional steel and aluminum alloys. It is as strong as steel, but 45 percent lighter. It is 60 percent heavier than aluminum, but twice as strong. Titanium is expensive only because the current process for refining the ore to metal is a multi-step, high temperature batch process. According to the AMPTIAC Quarterly, a Department of Defense-sponsored publication, current global production of titanium metal is approximately 50,000 tons per year at a market value of $600 million. AMPTIAC estimates that, due to the current state of manufacturing, titanium is produced at only about 1/20th of its current potential world volume. It is widely believed that a reduction of cost in the manufacturing process will expand the use to approximately one million tons. The new patent describes the process for making microporous structures that can be used as a catalyst support or for electrode material. The microporous structures have a high porosity and high thermal stability, combined with good mechanical strength and relatively high surface area. In addition to the structures being used in a process to make titanium metal, the process is useful for making titanium dioxide for catalyst structures used for fuel cells, sensors, and electrochemical cells.

Nano & Micro Electromechanical Systems

Corporation has announced their intention to enter the spectroscopy and applied spectroscopy markets by launching the OSM Series of turnkey spectrometers, light sources and fiber based accessories. These feature-rich instruments incorporate multi-element detector arrays, optional fiber coupling and battery-operated portability, as well as a choice of detectors and gratings covering the UV through near IR. Higher end, self-contained models provide innovative options such as on-board 12-bit processing for data acquisition and real-time data analysis, on-board memory and removable SmartMedia memory cards for data storage and transport, simple touch screen control, and network connectivity via an optional Ethernet port. The OSM series is targeted at demanding laboratory research applications, process monitoring and QC labs, as well as OEM system-integrators. Ron Hartmayer, Director of Marketing for Photonics Products, notes that, "Spectroscopic instrumentation is an obvious and natural progression for Newport on several fronts. First, we have long been supporting this market at the component level with both optics and high performance opto-mechanical assemblies. Second, all the core technologies deployed in these instruments are already strongly represented at Newport. Lastly, Newport has a history of supporting growing markets with value-added, integrated solutions that simplify the implementation of disparate photonic components. We certainly believe that spectroscopy and applied spectroscopy are markets with tremendous growth potential." Hartmayer lists an extensive range of applications that the company intends to actively support. These include basic chemistry and biochemistry, biology and biomedicine, semiconductor research and fabrication, pharma, astronomy, agricultural products and food processing, and education. More information on Newport's growing line of spectroscopy instrumentation is available on the company's Web site at www.newport.com.

Nanoelectronics

researchers Kathryn Guarini and Chuck Black made a splash on Dec. 9 at the IEEE's International Electron Devices Meeting in Washington, D.C., when they announced that they made a nanocrystal version of flash memory. But IBM is more bedazzled by the versatility of the fabrication technique than the device, according to Guarini. The flash application allowed them to prove the feasibility of their approach.

IBM uses a technique developed at the University of Massachusetts that allows polymers to self-assemble into a honeycomb pattern with holes as small as 20 nanometers. The polymers then are placed as a stencil on silicon dioxide, a material compatible with today's chip-making processes, for growing uniform nanocrystals. Guarini said her research as a graduate student at Stanford University, where she specialized in maskless lithography techniques, taught her how to build nanostructures using probe tips. The self-assembled honeycomb shape is one of several patterns Guarini said they want to create using various self-assembling materials. Different geometries would serve different functions as memory or logic devices. And their self-assembly system could complement other nanotechnology initiatives at IBM, such as efforts to construct transistors using carbon nanotubes. Placing nanotubes into desired positions is one of many challenges for making a nanotube-based transistor. A simpler and faster approach is to grow the nanotubes where they need to be. That may be possible by placing a catalyst into a honeycomb chamber from which the nanotube could form. IBM is in a race with companies such as Intel Corp., Hewlett-Packard Co. and Motorola Inc. to shrink components on chips for smaller, faster computing devices. The market for nanomaterials used in electronic devices will start in the billions and continue to grow, according to industry analysts at Business Communications Co.

Nanoelectronics

19, 2004 – When is a wire more than a wire? When you add the "nano" prefix, of course.

On one level, "nanowire" is a literal term: a channel for electrons or photons no wider than a few thousand atoms. As electronic components shrink closer to the size of biological cells or molecules, such wires may eventually need to replace the strands of metal that interconnect circuits today.

Nanowires are pinned under two electrodes.

Peidong Yang (News, Web) and colleagues at the University of California, Berkeley, have, for example, directed crystals of zinc oxide, a semiconductor, to self-assemble into arrays of nanowires that function as ultraminiature lasers and might be useful for probing cells or performing more-precise surgery. Yang's group has also put a thin layer of silver nanowires to work as a delicate chemical detector. Such a film of nanowires could be a good way to sense chemical or biological warfare agents. One of the most novel nanowires developed to date doesn't merely connect circuit components, but can work as an actual transistor or logic circuit. Charles Lieber's (News, Web) research group at Harvard fabricated such nanowires-as-devices by alternating bands of different semiconductor materials as the wire is formed. The nanowires can then be applied at room temperature to the back surface of a display as an extremely thin coating. The wires lie flat and all point in the same direction, a little akin to a raft of logs on a river. Next, a grid of electrodes is deposited using conventional manufacturing methods on top of the nanowire film. The electrodes pin bundles of the nanowires across a gap to form the transistor elements that make up a display's pixels. And because the process is low temperature, the nanowire film could be deposited on a flexible surface in a low-cost, high-throughput process such as "roll-to-roll," much the way newspaper is printed. On Jan. 14, Nanosys announced it would work with Intel to investigate how its nanotechnologies might work in future memory devices. The company, however, declined to specify whether such efforts would focus on nanowires.

Nanomedicine

Calif., Jan. 5, 2003- Researchers have developed an improved method for performing sentinel lymph node biopsy, a crucial first step in determining whether a cancer has spread to other parts of the body. The new method depends on quantum dots, nanometer-sized crystals that emit near-infrared light, to illuminate lymph nodes during cancer surgery. The research, resulting from collaboration between researchers at MIT, Beth Israel Deaconess Medical Center and Brigham and Women's Hospital, will be published in the January issue of Nature Biotechnology. The new near infra-red quantum dots were developed and synthesized at the MIT department of chemistry in the laboratory of Professor Moungi Bawendi, a scientific co-founder and advisory board member of Quantum Dot Corporation (QDC). The novel intraoperative, near-infrared fluorescence imaging system was developed in the laboratory of Dr. John Frangioni, Assistant Professor of Medicine and Radiology at Harvard Medical School and an Attending Physician at Beth Israel Deaconess Medical Center. The study, entitled "Near-Infrared Fluorescent Type-II Quantum Dots for Sentinel Lymph Node Mapping," by Sungjee Kim, and colleagues, describes how the quantum dots were injected into live pigs and followed visually to the lymph system just beneath the skin of the animals. The new imaging technique allowed the surgeons to clearly see the target lymph nodes without cutting the animals' skin. Sentinel Lymph Node (SLN) mapping, the surgical technique employed in the study, is a common procedure used to identify the presence of cancer in a single, "sentinel" lymph node, thus avoiding the removal of a patient's entire lymph system. SLN mapping relies on a combination of radioactivity and organic dyes but the technique is inexact during surgery, often leading to removal of much more of the lymph system than necessary, causing unwanted trauma. The current work was performed on laboratory animals, including near-human sized pigs, considered by scientists to be a good predictor of human results.

The study reported that the new imaging system with near-infrared quantum dots was a significant improvement over the dye/radioactivity method currently used to perform SLN for several reasons, including: Throughout the procedure, the quantum dots were clearly visible using the imaging system, allowing the surgeon to see not only the lymph nodes, but also the underlying anatomy.

The imaging system and quantum dots allowed the pathologist to focus on specific parts of the SLN that would be most likely to contain malignant cells, if cancer were present. The imaging system and quantum dots minimized inaccuracies and permitted real-time confirmation of the total removal of the target lymph nodes, drastically reducing the potential for repeated procedures. "SLN mapping has already revolutionized cancer surgery. Near-infrared quantum dots have the potential to improve this important technique even further," said Dr. Frangioni. "This is an impressive study," added Carol Lou, president of QDC. "While still several years away from being a reality for patients, this is a wonderful demonstration of the potential of quantum dot nanotechnology to significantly enhance medical care." The paper can be found on the Nature Biotechnology website at: www.nature.com/cgi-taf/dynapage.taf?file=/nbt/journal/vaop/ncurrent/index.html

Nanomaterials

working to make microscopic and nanoscale machines and electronics have produced electrical wires at that scale, but it has proved more difficult to shrink the fiber optics that guide light. The trick to guiding light is finding ways to keep the photons confined to a fiber rather than leaking out. Researchers from Harvard University, Zhejiang University in China and Tohoku University in Japan have made glass optical fibers as thin as 50 nanometers that guide light without losing much of it. Fifty nanometers is more than one thousand times finer than human hair. The researchers have made the thin optical fibers up to several centimeters long. The key to such small, low-loss optical fibers was finding a way to make them very uniform in diameter and with very smooth walls, according to the researchers. To make the fibers the researchers first used a flame-drawing method to make micron-sized fiber. They then wound the fiber around a tapered, heated, 80-micron-diameter sapphire tip to keep the wire at a steady temperature while they pulled the fiber a second time to make it thinner.

The tiny optical fibers could be used in microphotonic devices for optical communications and optical sensing. The smaller fibers could lead to smaller and/or faster devices, according to the researchers.

Nanomaterials

Nanomaterials are well known but the development of a new class of plastics that can be reshaped under pressure could lead to a new approach to manufacturing plastic products that avoids energy-guzzling high-temperature processing. US researchers devised the "ever green" plastics, which are also recyclable materials, to maintain their useful mechanical properties even after they have been remoulded.

Anne Mayes and her colleagues in the Department of Materials Science at MIT (Massachusetts Institute of Technology) in Cambridge have made the new class of plastics—baroplastics—from a nanophase blend of two polymers, polystyrene and poly(n-butyl acrylate) or polystyrene and poly(2-ethylhexyl acrylate). Under several hundred times atmospheric pressure, at 34.5 MPa, the clear solid softens and can be moulded into any shape desired. The idea of using raised pressure rather than temperature to process two otherwise immiscible polymer components is supported by small-angle neutron scattering (SANS) carried out by the team.

SANS reveals how low-temperature processing affects morphology too. Samples before and after room-temperature processing were analysed as well as after ten "recycling" cycles. The SANS measurements revealed how the rigid PS and soft PEHA mix together during processing, enabling the material to flow.

An additional advantage of using high-pressures as opposed to high-temperature for plastics processing is that the plastics themselves are not degraded by pressure in the way that they are by heating them even just to 200 Celsius. This means that the Mayes' nanophase polymer blends are almost limitless in the number of times they can be recycled without degradation. The team shredded and recycled their two baroplastics ten times and observed no obvious changes in molecular properties of the product as determined by preliminary mechanical and optical property testing and differential scanning calorimetry. Degradation is one of the serious limiting factors in current plastics recycling as it can destroy flame retardants and ultraviolet protectant additives as well as the polymers themselves. "The baroplastics could provide alternative materials for applications currently employing semicrystalline polymers, rubber-modified plastics or thermoplastic elastomers, or as a matrix material for other thermally sensitive components such as biologically-derived therapeutics," Mayes told Spectral Lines. "The new materials should be compatible with much of the plastics-manufacturing equipment already in use," she adds, "which generally employ pressures of magnitudes comparable to those used in the present study."

Advanced Research Laboratory (ARL) is getting ready to commercialize a low-cost "nanostamp" technology for medical applications. Hitachi's process creates "nanopillars" with extremely high aspect ratios (narrow relative to height), a feature that the company believes will prove useful for biochips and other applications, according to Akihiro Miyauchi, a senior researcher at Hitachi.

The technology uses a silicon "stamp" that presses onto a polystyrene-based polymer film, producing nanopillars that are extremely long and thin, about 3 microns in height. Right now, ARL is concentrating on making two versions of the nanopillars: one that is 250 nanometers in diameter and another that's 80 nanometers.

"With the 80-nanometer-diameter nanopillars, we are already producing pillars that are smaller than the current (90 nanometer) semiconductor process node by Intel or Fujitsu, for example, but we can easily go smaller because we have a very simple and very cheap process; it's just press and release," Miyauchi said. He did not say how much smaller ARL plans to make the nanopillars, but he did say the process already works out to be much more cost-effective than semiconductor lithography processes. Hitachi is not alone in developing very high-aspect ratio and market-ready process, according to Lars Samuelson, leader of the Nanometer Consortium at Lund University in Sweden. Lund's department recently developed techniques to direct self-assembly of nanowhiskers, he said. This nano-imprint lithography (NIL) technique can make stamps with features of less than 20 nanometers in diameter, and "maybe close to 1: 100 in aspect ratio or even larger when optimized," Samuelson said. Hitachi, meanwhile, is also gearing up for commercialization. Having developed the production technology, Hitachi has created a 15-member nanotechnology business enhancement office. Hiroshi Sonoda, senior engineer at Hitachi's nanotechnology planning office, said there is a $35 million market for the application in Japan alone after full-scale production. Hitachi said it plans to start shipping samples in the near future.

Neil Gordon, a partner at nanotech consultancy Sygertech and president of the Canadian NanoBusiness Alliance,said that development of low-cost products is key to applications like disposable medical tests. Hitachi, being a big multinational company, can afford to subsidize its nanopatterning products indefinitely, he said, giving the company "a considerable competitive advantage over dedicated nano-imprint lithography vendors." To facilitate physical communication with the external environment, the vasculoid may be equipped with mechanical external ports which permit the ready attachment of cables or peripheral equipment as an alternative to purely wireless (e.g., radio, optical) communication links. Vasculoid ports include an appropriate macroscopic connector receptacle in hard contact with the sapphire structure. Ports may include connections to the internal plate-to-plate acoustic communications network or to docking bays and cellulocks to permit access to the internal material flow and thus allow inserting or extracting: (1) molecules and cells; (2) fresh vasculocytes and spare parts; (3) bioprovisions including respiratory gases, water and energy supplies; and (4) nanomechanical or other waste material. Extracorporeal user control interfaces could also be connected via an external port, if needed. Convenient and aesthetic locations for external ports include the navel or the nape of the neck. Respiratory gas and glucose transport requirements could be reduced by importing externally-supplied electrical energy—e.g., wall sockets (110 VAC at 0.9 amps equals the 100 watt human basal rate), backpack generators, solar collectors, nuclear batteries etc.—to be distributed throughout the appliance via wiring in the plates. These sources could be used to power: (1) local recycling of CO_2 and H_2O waste back into O_2 and high-energy carbohydrate fuel (mimicking photosynthesis energized by light), (2) the reconstitution of amino acids and proteins from urea and other nitrogenous wastes (as found in ruminants and in hibernating bears), and (3) other "reversible nutrition" processes, converting the body of the envasculoided user into a more closed-cycle system with reduced dependence on certain material and energy inputs. However, this approach would correspondingly increase user dependence on the chosen external power source and would require new in-plate or in-vesicle chemical processing plants without completely eliminating the need for any existing subsystem, hence may not be worth the added complexity.

Vasculoid installation leaves the user's gross thermal mass largely unchanged, as only ~4.4 kg of blood borne water (~9 per cent of the ~50 kg normally present in the human body) is removed and replaced with ~2 kg of sapphire vasculoid with a heat capacity equivalent to ~0.2 kg of water. Total passive aqueous heat capacity drops from 50.4 kcal/K down to 46.3 kcal/K. Nevertheless, 20th century patients with extensive metallic implants (e.g. pins, plates, bolts and joints) occasionally report brief chilly sensations during periods of cold weather and in other circumstances, due to the high thermal conductivity of metal compared with natural biomaterials or with plastics. Some organs such as the cornea are quite sensitive to sudden temperature changes—as little as a 0.3 K drop over a 0.785 mm^2 area for a duration of 0.9 sec (~700 nanojoules) is detectable by patients. A thermally conductive sapphire or diamondoid-coated sapphire vascular implant could extend these unpleasant sensations throughout the entire body, even for activities as simple as manually grasping a cold object. A targeted thermogenerative subsystem could help to eliminate this effect.

Perspiratory thermoregulation will continue as before. However, as explained earlier the active heat transfer mechanism of blood will be completely disabled. Normal capillary vasoconstriction/vasodilation mechanisms may be at least partially suppressed to avoid unnecessary tensions at the vasculoid-vessel interface. Detection of these responses, or direct temperature measurement, may be used to adjust passive thermal conductivity over a wide range, as follows. Each plate abuts neighboring plates through metamorphic bumpers with optimal thermal contact along stripe-like diamondoid buttons. Within each bumper, we may place an opposed pair of diamondoid pistons, separated by vacuum when they are pulled apart. When pulled apart, heat can be conducted only through the sapphire infrastructure, or vacuum in the plate, or through a surrounding aqueous medium, and is thus very slow, almost the same as normal tissue. But when the two pistons are pressed tightly together against diamondoid bumper buttons that are in good thermal contact, heat conduction largely bypasses the poorly-conducting sapphire regions and flows almost exclusively through the diamondoid contact region. Thus with pistons apart, the envasculoided human body has near-normal thermal conductivity; with pistons in contact, the body's thermal conductivity becomes near-metallic, perhaps as conductive as stainless steel. This transition is subject to user or program control, is switchable in microseconds, and may be directed only to specific volumes or pathways within the body if so desired*. A more complex system might employ thermal rectifiers, or, as J.S. Hall suggests, "heat pumps could be placed in the joints to make the whole phenomenon usefully controllable, augmented if necessary with tankers of ice and/or steam."

Installation of a full vasculoid appliance permanently displaces ~4.2 liters of natural blood volume, freeing up ~1 gallon of internal storage volume. At least some of this volume may be used for containerized temporary caching of consumables including surplus glucose, fats, water, oxygen (e.g. a high-pressure nanolung), minerals or other useful biomaterials, or various bio- and nano-wastes. For example, a 1-liter 1000-atm O_2 solid-walled cache holds ~34 hours of oxygen at the human basal metabolic rate or ~2 hours at the maximum rate. The free ~4.2-liter volume could also be used to store a wide variety of useful equipment or tools including computers, computer memories, external communications or navigational devices, solar energy accumulators, weapons, spare vasculocytes and other special purpose nanorobots, spare parts, or useful tools. An entire spare vasculoid (~558 cm^3) could even be stored in this volume.

The vasculoid, like the micron-sized respirocytes proposed elsewhere, will offer the ability to breathe at low O_2 partial pressures. The vasculoid gas tanker fleet can hold ~20 minutes of oxygen at the basal metabolic usage rate, or up to ~100 minutes if most of the tanker fleet is diverted to respiratory gas carriage in lieu of other applications. Vasculoid installed in underwater divers who are breathing pressurized air to modest depths should allow only enough nitrogen into the body to forestall mechanical tissue damage, then rotor it back out again as the diver surfaces to avoid decompression sickness. At 100 per cent saturation the body absorbs ~10^{21} excess N_2 molecules per meter of depth. Each tanker could hold ~7.98 × 10^9 nitrogen molecules (at 1000 atm), requiring the storage capacity of 0.125 trillion tankers per each 1 meter of decompression. Given a reconfigurable fleet of ~166 trillion tankers, expedited decompressions from ~100 meters are probably achievable using the present design. To establish a greater vasculoid operating depth, in theory an auxiliary 1000-atm storage tank "nanolung" installed in the vasculoid wall could provide ~9.43 cm^3 N_2 storage capacity per each 100 meters of decompressible diving depth. However, nitrogen should not be used to hyperpressurize the tissues due to nitrogen narcosis. It should also be noted that obesity is a risk factor in decompression sickness because nitrogen is 5 times more soluble in fat than in water and because of reduced blood access to adipose tissue, and also compression and decompression are asymmetrical: a nitrogen load acquired in a few minutes may take hours to fully deplete purely by diffusion (and the extraction rate varies markedly by tissue type), so extravasculoid respirocyte-class nanorobots may be required for optimal results. Clearly the vasculoid can achieve a decompression rate comparable to breathing pure oxygen, but without the oxygen toxicity. Helium may be useful for mixed-gas diving, but is prone to leakage some will be lost from the body during use and cannot effectively be replaced from the environment.

The lymphatic system extends into all the same tissues as capillaries and is structurally similar to the venule network. However, since lymph vessels constitute a "cul-de-sac" system, not a "circulatory" system, it is unnecessary to install vasculoid systems in the lymphatics in order to achieve most of the positive benefits expected from an arteriovenous vasculoid. With the circulatory vasculoid in place, the workload of the natural lymphatic system may be greatly reduced—protein leakage from capillaries, entry of particulates into the tissues, and the presence of pathogens requiring immune system response all should be greatly reduced. More properly, lymphovasculoid should be regarded as optional equipment which may permit more precise control of the immune system in general and of lymphocyte traffic in particular. Total adult lymph flow is typically ~2 liters/day, so transport requirements within the lymphovasculoid should not be particularly severe. If implemented, the lymphovasculoid would be emplaced as two distinct installations—a right lymphatic and a thoracic subsystem—following the natural division in the human body. Special lymph-recycling facilities may be required at the locations where the thoracic duct and the right lymphatic duct join the venous tree.

Pathogen disposal sites using an enzyme-based digest and discharge protocol may be located near lymphoid tissues and organs (to permit immune system processing) and near excretory organs such as kidney, liver, gallbladder, and gut (to facilitate removal from the body). Specialization of this function at specific locations appears more efficient than on-site pathogen processing systems which would need to be numerous, widely distributed, and usually idle.

It may be useful to deploy sensor probes which can leave the vasculoid and enter the surrounding tissues to detect and monitor remote events (such as angiogenesis, tumors, or pathogens) that otherwise might not be conveniently detected from within the vasculoid. Additionally, large blocks of tissue are only lightly vascularized or have no capillaries whatsoever, as for instance the synovial chambers in skeletal joints and the epidermis. Medical nanorobots capable of migrating through noncellular tissue might be useful both for repair purposes and for maintaining a disease-free condition in these tissues. Extravasculoid devices may also constitute general-purpose mobile cell repair machines with even broader injury response and prevention capabilities.

Although the vasculoid itself is essentially fireproof (sapphire and ruby will not burn in oxygen), the overlying human tissue is not. Thus another optional upgrade to the basic vasculoid package is an active damage control subsystem that offers some protection from severe thermal burns. This optional subsystem would include installation of biocompatible sensor-tipped diamondoid or sapphire pores that pass from the epidermis through the dermis to the nearest envasculoided capillaries. The comparative human skin response to sapphire vs. fluid cooling has been studied experimentally. Upon detection of a potentially harmful thermal event, sacrificial water is pumped from internal reservoirs into channels in the plates, and from there into the pores, eventually emerging as billions of aqueous microstreams. Touching a red hot object stimulates a nearly instantaneous emission of ablative water, producing a protective layer of heated water and steam between the hot object and the skin. Protection of the human hand (~150 cm^2) up to ~1300 K (~decomposition temperature of diamond) for a ~1 second exposure requires a 162 kilowatt/m^2 heat dissipation rate and a ~1 cm^3 sacrificial water reservoir. Of course, this system could easily be overwhelmed if the user is exposed to intense IR radiation, such as inside burning buildings or near large explosions, and in any case the hair is not protected as well as the skin.

The overall static structure of the basic vasculoid appliance is an array of rigid plates, strongly fastened together, entirely covering a curved two-dimensional surface embedded in a three-dimensional space. The plates conform to ordinary movement, but may adequately resist certain kinds of destructive movement such as separation due to cutting, rapid accelerations or decelerations, and crushing injuries. Although local endothelial cells may be damaged during such sharp movements, an envasculoided tissue should be somewhat more resistant to gross damage. For example, the critical buckling pressure of a hollow tube of circular cross-section deformed into an elliptical cylinder by external pressure is given by Freitas as $p_{crit} = (E\ h_{tube}{}^3)/(4\ r_{tube}{}^3\ (1 - c_{Poisson}{}^2))$ where E is Young's modulus (~10^6 N/m^2 for vascular tissue and ~10^{11} N/m^2 for sapphire, h_{tube} is tube wall thickness (~1 mm for aorta, ~ 1 mm for capillary and for vasculoid plate-tube, r_{tube} is tube inner radius (~25.0 mm for aorta, ~8 mm for capillary, ~7 mm for vasculoid plate-tube; and $c_{Poisson}$ is the Poisson ratio for the material (~0.3 for vascular tissue, ~0.1 for diamondoid). Using this formula and the stated values, p_{crit} for aorta is ~59 N/m^2 and the vasculoid coating adds only negligible additional resistance to crushing, p_{crit} ~ 0.002 N/m^2. However, for the ubiquitous capillaries p_{crit} ~ 540 N/m^2 but adding the vasculoid coating can dramatically increase crushing resistance up to ~7 x 10^7 N/m^2, or ~700 atm of overpressure. Since the mechanical coupling between plate bumpers is subject to both design and conscious user control the crushing resistance of capillary beds can be autogenously varied over five orders of magnitude. Of course, high overpressures may seriously damage the underlying natural endothelium but this might nevertheless be acceptable over small areas during emergency situations. For instance, resistance to deep incision-slash wounds would be markedly improved. Crushing strength may be increased using thicker plates or stronger bumpers.

Similarly, the bending stiffness of a hollow tube is given by Freitas as $k_{shaft} = 3p\ E\ (R_{tube}{}^4 - r_{tube}{}^4) / (4\ L_{tube}{}^3)$, where the outside tube radius is $R_{tube} = r_{tube} + h_{tube}$ and the tube length is L_{tube} (~400 mm for aorta, ~1 mm for capillary. Using this formula and the stated values, the resistance to bending (stiffness) of the aorta increases 60-fold (to ~230 N/m) when coated with a single layer of vasculoid plates, but the stiffness of capillaries increases nearly 70,000-fold (to ~0.4 N/m) when coated with diamondoid plates. The buckling strength of coated vessels compared to natural vessels may exhibit similar orders-of-magnitude improvement, though in practical systems some of this potential will be lost because interbumper connection rupture strength may be only 10^7-10^9 N/m^2, 1-3 orders of magnitude less than for solid diamondoid materials. The natural musculature may be too weak to move soft tissues that are thoroughly penetrated with ultrastiff capillaries, but selective autogenous control of plate bumper expansion and contraction should make possible any necessary macroscale voluntary movements, or on-demand mechanical rigidification of specific limbs or organs. Bumpers may be driven at ~KHz frequencies, permitting positional adjustments on millisecond timescales. Such motions will require additional energy expenditure.

A complete examination of the acceleration tolerance of the envasculoided whole human body would have to take into account a wide variety of factors including differential density of body parts (e.g., lung, soft tissue, skeleton), magnitude and direction of the stress vector (e.g., positive or negative, axial or transverse, etc.), duration and timing of the stress vector (e.g., acute or chronic, linear or periodic), rate of onset and pulse shape of the acceleration, and the mechanical characteristics both of linked vasculoid components and of the vasculoid-endothelial interface. However, a simple order-of-magnitude estimate may be cautiously ventured, as follows. A vasculoid appliance of differential density Dr_{vasc} relative to soft tissue and mean thickness h_{vasc} in contact with biological tissues that are subjected to a linear acceleration of $a = (a_G\ g)$, where $g = 9.81\ m/sec^2$, pushes the vasculoid into the tissues with a pressure force of $P_{vasc} \sim h_{vasc}\ Dr_{vasc}\ g\ a_G$. Because the vasculoid tube has 2 opposed walls, each 1 mm thick, and because the tanker fleet covers 4 per cent-55 per cent of the surface of each of these opposed walls with tankers that are 1 mm thick, then the effective capillary thickness of pushed vasculoid under acceleration ranges from $h_{vasc} \sim 2.1\text{-}3.1$ microns. Tissue density is $r_{tiss} \sim 1050\ kg/m^3$; sapphire tanker density is $r_{tank} \sim 993\text{-}1738\ kg/m^3$ if filled with vacuum or water at 310 K, and plate density is $r_{plate} \sim 2000\ kg/m^3$, giving an effective vasculoid density of $r_{vasc} \sim 1643\text{-}1988\ kg/m^3$ under various usage conditions. Hence $Dr_{vasc} = r_{vasc} — r_{tiss} \sim 593\text{-}938\ kg/m^3$ and so the maximum tolerable G-force for envasculoided tissue is $a_G \sim P_{vasc} / (h_{vasc}\ Dr_{vasc}\ g) = 24.4\ P_{vasc} \sim 4880$ G, taking $P_{vasc} \sim 200\ N/m^2$ as the tentative limit for acute vascular damage because this has been found experimentally not to cause gross injury or denudation of canine arterial endothelium by shear force.

However, endothelial transcription factor responses (protein expression) can be triggered by shear forces as low as $0.02\text{-}2\ N/m^2$ ($a_G \sim 0.5\text{-}48.8$ G) and normal physiological blood shear forces are $0.14\text{-}2.6\ N/m^2$ ($a_G \sim 3.4\text{-}63.4$ G), so the maximum chronic external acceleration indefinitely tolerable by an envasculoided human body probably may not exceed 50-100 G without producing altered, possibly pathological, cytochemical states. (The operational limit of the current vasculoid design is also ~30-100 G). This still would represent a substantial improvement over the natural acceleration tolerance of the unaided human body—the mean relaxed tolerance is ~3.23 G and loss of brain blood flow occurs at $+G_z > 4.5$ G. Consciousness has been retained for a maximum of 45 seconds at 9 G using G-protective equipment and straining maneuvers developed for the U.S. Air Force, and for 4 minutes at 12 G or 4 seconds at ~16 G using water immersion. Severe impact injury to humans occurs from $+G_z$ accelerations of 30 G exceeding 100 millisec or 100 G exceeding 2 millisec—the recordholding primate is evidently one of Colonel Stapp's chimpanzee test subjects that survived ~1 millisec of 247 G in the $-G_x$ direction on a rocket sled, suffering only "moderate" injuries. Thus, the basic vasculoid may improve maximum human acceleration tolerance by a factor of 10-20, or more, and higher-G-tolerant architectures can probably be devised. With sufficient capability, the vasculoid structure could allow the non-local distribution of stress—for example, by spreading the force of an impact in a manner similar to a bulletproof vest, also helping to quickly close any vascular breaches that might occur. A common and significant personal injury mechanism is brain-skull impact, and a tough interwoven vascular support scaffolding could help to fix the brain within the skull, potentially reducing the danger of concussion for at least moderate decelerative loads. (Active suppression of rotational impact trajectories may enhance cerebral durability, as demonstrated using high-speed cinematograph films of woodpeckers which regularly survive 6-7 m/sec cranial impact velocities with ~1000 G decelerations.) Positional information and selective sectioning could allow unavoidable partitions of tissue to occur along straight planes, producing a minimum of damage when compared with ordinary tearing injuries. Whether the vasculoid appliance would be a liability in explosive overpressure situations, where a fluid-filled vasculature might not (e.g., lethal effects in humans noted for 40 psi to 50 psi overpressure shockwave; 300 psi (~20 atm) record for successful deep water submarine escape), deserves further study. A preliminary scaling study for a conceptual design of a single, complex, multisegmented nanotechnological robot that appears capable—with numerous caveats, as noted—of duplicating all essential thermal and biochemical transport functions of the blood, including circulation of respiratory gases, glucose, specialty biochemicals, waste products, and all blood borne cellular elements. The vasculoid, a 2-kg ~200-watt intimate personal appliance, conforms to the shape of the existing vasculature and may serve as a complete replacement for natural blood while greatly improving the durability and functionality of the human body. The device described in these chapters would represent a most extreme intervention using a very advanced medical molecular nanotechnology. It also should be noted that our current knowledge of the biological functions of the circulatory system is incomplete, so the design presented here must be considered provisional at best. But the principal challenge was to advance a plausible argument that a nanomechanical whole-body thermal and biomaterials transport system would violate no known physical, engineering, or medical principles, could presumptively be made adequately safe for the user, and might confer some significant advantages over simpler whole-body systems exclusively employing unlinked populations of individual blood borne and tissue-borne nanorobots.

Ultimately, and from the standpoint of human-guided evolution, the body exists primarily to ensure the survival of the mind—not the replication of the genes, which was the ancient paradigm. It would seem that a somewhat more advanced and compact version of the proposed device could function independently of nearly all noncortical tissue. Thus the vasculoid is most fascinating because it may represent one last outpost of humanity at the final frontier of biological evolution.

Applications and Importance Nano-Tribology

The term tribology is derived from the Greek word *tribo* meaning rubbing and *logy* meaning knowledge. The original applications by the Greeks of tribology were in trying to understand the motion of large stones across the earth's surface. Today tribology has grown to include the methodical study of friction, lubrication, and wear.

Tribology plays a critical role in diverse technological areas. In the advanced technological industries of semiconductor and data storage, tribological studies help optimize polishing processes and lubrication of data storage substrates. In traditional industries such as automotive and aerospace, tribological studies help increase the lifespan of mechanical components. Many industrial processes require a detailed understanding of tribology at the nanometer scale. The development of lubricants in the automobile industry depends on the adhesion of nanometer layers (mono layers) to a material surface. Assembly of components can depend critically on the adhesion of materials at the nanometer length scale.

There are a number of traditional tools for characterizing friction, lubrication and wear. The most common characterization tool is the tribometer having several configurations such as pin-on-disk, ball on flat, and flat on flat, etc. Generating motions at the nanometer scale is extremely challenging. New characterization techniques are required to understand tribology at the nanometer scale. The atomic force microscope is now being routinely applied for studying nanoscale tribology. The natural extension of the AFM for tribology applications is derived from the motion of a nanometer-sized stylus in the AFM over a surface. Although traditional tribology testing is not done with an AFM, many new types of applications are possible.

Examples of the application of AFM to tribology include:

- Direct three-dimensional visualization of wear tracks, or scars on a surface.
- Measurement of the thickness of solid and liquid lubricants having nanometer or even monolayer thickness.
- Measurement of frictional forces at the nanometer scale.
- Surface characterization of morphology, texture, and roughness.
- Evaluation of mechanical properties such as hardness and elasticity, and plastic deformation at the nanometer scale.

A major advantage of the AFM for tribological studies is that the AFM can be routinely used on all types of materials. Materials commonly studied include: ceramics, metals, polymers, semiconductors, magnetic, optical, and biomaterials. AFM investigations are usually made in ambient air environment. It is possible to make AFM studies in a vacuum or liquid environment. The effects of wear at the nanometer scale become critical to the optimization and stability of machines as the tolerances in precision machines become smaller and smaller. Traditional microscopes such as the optical and scanning electron microscopes facilitate visualization of wear in 2-dimensions. For example, with the SEM it is possible to get a magnified view of wear tracks in the x-y axis but cross sectioning is required for measuring the depth of wear tracks. The AFM allows direct 3-dimensional visualization of wear tracks and scars. The images may be displayed in a 2-D projection and a 3-D projection. Direct measure of wear track depth can be easily measured with a line profile derived from the AFM image.

It is well known that layers of lubricants on surfaces that are less than 100 nm can dramatically affect lubrication behavior. Characterization of such films is necessary for developing optimized lubricating films. However, nanometer scale characterization of lubrication films offers a substantial challenge. Optical techniques such ellipsometers can be used for measuring lubrication thickness of large sections, (greater than 10 square micrometers), of a surface. Measurement of the localized (less than 1 micron) film thickness is not possible with the ellipsometer. The probe is mounted at the end of a cantilever in an AFM making it possible to measure interaction forces between the probe and the surface by monitoring the deflection of the cantilever. A graph, called a force/distance curve, shows the forces on the probe as the distance between the probe and the surface are reduced. The nature of the force/distance curve depends on the force constant of the cantilever, the lubrication density, probe geometry, and the lubrication thickness. By measuring the changes in force/distance curves in an AFM it is possible to directly ascertain the thickness of lubrication films. Below is an example of a force/distance curve for a surface with no lubrication film compared to one with a lubrication film. The thickness of the film is established from the force/distance curve.

Friction between two surfaces depends on the chemical and mechanical interaction between the surfaces. Changes in chemical composition giving rise to friction are measurable with the AFM. The technique for measuring these forces is called lateral force, or frictional force microscopy. As the probe moves over a surface in the AFM, changes in the chemical composition of the surface can give rise to torsions of the cantilever on which the probe is mounted. The torsion of the cantilever is then proportional to the friction between the probe and the surface. In an AFM it is possible to simultaneously measure topography and frictional force images. The topography image is derived from monitoring the vertical forces on the cantilever and the friction image is acquired simultaneously by monitoring the lateral motions of the cantilever. Below is a FFM image of a sample illustrating changes in the friction.

The AFM gives extremely high contrast on surfaces that are flat at the nanometer scale. Optical and electron microscopes are not able to resolve surface texture that is easily measured with the AFM. Applications include the visualization of surface topography in 2-D and 3-D perspectives, line roughness measurements, and area roughness measurements. All of the traditional area and surface roughness parameters can be calculated after the AFM image is acquired.

Mechanical properties such as hardness, elastic modulus, stiffness and compressibility as well as material behavior such as plastic deformation, and fracture can be studied with the AFM. It is possible to study nanohardness by directly pressing an AFM probe into a sample's surface; however, it is advantageous to use an instrument that is optimized for nanoindentation. The primary advantage of the nanoindenter over an AFM for nanohardness measurements is that it is easier to get calibrated measurements with the nanoindenter. It is useful to use the AFM to measure the three-dimensional topography of indentations made with a nanoindenter. AFM images allow direct visualization of material deformation or fracture behavior.

Using techniques such as pulsed force mode, the stiffness of a sample at a matrix of locations is measurable. From this data it is possible to create a stiffness mapping of a surface. Stiffness maps can only be made on samples where the stiffness of the surface is lower than the stiffness of the cantilever. Such stiffness images are routinely measured on polymer samples. Adding a fixture to the stage of the AFM makes the study of material behavior such as plastic deformation and fracture possible. The fixture permits creating forces on a sample while AFM images are being taken. A variety of materials may be studied with such a technique.

Tribiology (friction, lubrication, and wear) is the science of interacting surfaces in relative motion. The word "tribology" is from the Greek *tribein*, meaning to rub. Although the term may be unfamiliar to many, its meaning is clearly conveyed among some of the oldest written records. An Egyptian painting dating back to 1880 BC depicts workers dragging a sled containing a heavy statue. One worker pours a liquid on the ground just before the runners to make the going easier. Leonardo da Vinci addressed the same problem when he wrote in 1519, "All things and everything whatsoever however thin it be, which interposed in the middle between objects that rub together, lighten the difficulty of this friction." An excellent source for further historical reading is *Dowson's History of Tribology*. We have first-hand experience with friction any time we walk on ordinary ground and compare the outcome with what happens when we try to walk on ice. Wear of mechanical parts—as every car owner knows—eventually leads to failure and is one of the most costly problems facing industry.

To design and engineer mechanical parts, we need a macroscopic understanding of tribological processes. However, today the tolerances of many components, such as optical devices, computer chips, and ultra-smooth surfaces, are approaching the microscopic (atomic) length scale. In fabricating such precise components, we can greatly benefit from an atomic scale understanding of what happens when surfaces interact. The emerging field that lets us obtain this atomic scale understanding of fundamental processes of surfaces in motion is called molecular tribology or nanotribology. (A nanometer is one billionth or 10Ð9 of a meter; the size of an atom is about 0.3 nm.) Recent progress in nanotribology is being driven by advances on several fronts. Of primary importance is the advent of probes that allow researchers to examine interacting surfaces at a microscopic level. One such probe is the atomic-force microscope, which has a very small tip with a radius of about 10 nm used to probe a surface. The surface force apparatus has also become a highly useful tool. Other major developments in the last decade are better models of interatomic forces, which, coupled with improved performance of computers, are giving researchers a new window into tip—surface interactions through computer simulations.

The molecular dynamics method was originally developed in the late 1950s to study the statistical mechanical properties of a collection of atoms at equilibrium. The method is now a well-established and important tool in the fields of physics, chemistry, biology, and engineering. Recent advances in computer hardware and software allow us to simulate relatively large systems using increasingly realistic models of the forces between atoms.

In principle, MD modeling is simple. We start by identifying the position and velocities of all atoms in a system we want to model. Then we calculate the force on every atom from its neighbors and advance the positions of the modeled atoms according to Newton's equations of motion (force = mass × acceleration). During a simulation, we evaluate the response of a material—often copper, silicon, or silica glass—as it is subjected to an external force by following the response of each atom. Examples of external forces include stress fields, moving boundaries, and heat baths. Despite the simplicity of MD modeling, the number of users and applications today remains limited; in addition, many researchers do not use it to its full potential. The reason for this under use is that this type of modeling can be "overkill" in that it is too expensive and requires too much detail for solving most practical problems. Furthermore, there is a certain art in selecting the appropriate boundary conditions and constructing interatomic-force models.

The boundary conditions used in many of the simulations of indentation and cutting of a clean metal surface by a diamond-like tool. In this figure, the dashed line forms a box surrounding all the atoms that are represented. This box is called a simulation cell, which is the "window" on the system being modeled. Both the metal surface and the V-shaped diamond tool above it have "handles" to position the surfaces relative to one another. These are the dark circles forming the boundary. In practice, the tool is fixed in the rotational direction and slide the work surface from left to right during a simulation of cutting. Immediately next to the boundary atoms is a region in which the atoms are constrained to be at room temperature. From an atomic point of view, temperature is a manifestation of the vibrational energy of the atoms. The role of the thermostat atoms in the tribology simulations is to draw away heat generated at the tip of the tool, mimicking a much larger piece of material than is represented in the illustration. The remaining atoms represented by open circles make up the Newtonian region, meaning that these atoms are free from further constraint and move according to Newton's equations. The tribological processes in this region are of interest in the MD simulation.

In many cases, we would like to simulate cutting or scraping over lengths that are far greater than the dimensions of the dozen or so atoms. However, as mentioned earlier, following the motion of a large number of atoms is prohibitive in terms of computer time and cost. To help overcome this problem, a relatively simple approach is used. Atoms are allowed to leave the simulation cell at the "downstream" end (to the right as the simulated work piece moves under the tool). At the same time, periodically new atoms are inserted from the left at the upstream boundary. This approach restricts the population of atoms simulated at any given time, while allowing to model nanotribological processes in a much larger work piece. The extension of this model from two to three dimensions is straightforward.

One of the first simulations we refer to here involved the indentation of a metal surface by a blunted, triangular diamond tip moving vertically at a constant velocity ranging from 1 to 1000 m/s. Keep in mind that in this and all the subsequent work talked about here, the objective is to understand—at a more microscopic, atomistic level than was previously possible—what is happening when surfaces in relative motion interact with one another. A "snapshot" of what happens when the tool indents the first three layers of metal. The interatomic forces in a metal arise from two effects: the interaction between electrons localized on the atoms, and the interaction of atoms with free electrons. An embedded-atom model of the metal is used that incorporates both of these effects. In this figure, the atoms are shaded by the local value of stress. At first, the surface responds elastically (deforms without permanent loss of size or shape), and we see circular regions of constant stress, called the Hertzian stress field, which is well known from elasticity studies of contact mechanics. When the tool is pushed in further to indent six layers of metal, the surface begins to flow plastically, and we obtain shear and discontinuity. In a 2D simulation like this one, dislocation edges are clearly visible within the metal along the darker colored slip bands that appear at 60° angles with the surface. In a close-packed metal such as this, the atoms behave like hard spheres (for example, billiard balls), and slip is analogous to sliding stacks of billiard balls over each other A 3D simulation of the indentation of copper gives a slightly different picture. An image obtained after copper was indented at a rate of 1 m/s, and the simulated tool was withdrawn from the surface. (Slower rates, typical of experiments performed in a laboratory, are beyond the current capabilities of MD simulations.) In a 3D simulation, we do not see the distinctive bands of dislocation that are observed in a 2D simulation. Moreover, there is relatively little distortion at the metal surface. Instead, we find a small pileup of atoms on the metal surface, a few atoms in interstitial positions (between regular lattice positions), and a small dislocation loop.

To explain the difference between the 2D and 3D simulations. Here, the atoms are colored according to the initial layering before the computer experiment. This cross section at the moment of full indentation of the copper surface shows that very little surface distortion has actually occurred. Remarkably few atoms have bulged out around the tool. Instead of distinctive dislocation bands, we see disorder only within a few layers of the sides and bottom of the tool tip. This localized disorder arises from point mechanisms (interstitials, vacancies, and so forth) of plasticity.

There are many other ways to display the results for such a simulation. One of the best and most revealing is by means of a graph, called a loading curve. The load (instantaneous force in nanoNewtons) on the diamond-like tool as a function of depth of indentation (shown here as the number of layers indented) for the simulation. The very rapid fluctuations in load—which look like "noise" on the curve—arise from the rapid motion of surface atoms repeatedly colliding against the tool atoms. The larger peaks and valleys are far more interesting in terms of what they reveal. At first, the load rises linearly as the metal surface responds elastically. After indenting about 1.5 layers of copper, the load drops abruptly when the elastic stress is relieved. This abrupt drop, called "critical yielding," is reminiscent of what is observed in the laboratory. The first yielding in the simulation corresponds to a single copper atom popping out onto the surface from under the tool tip and relieving the stress energy. With more indentation, more of these point events take place. After indenting seven layers of copper, the tool is stopped. At this point, the surface accommodates nearly all the tip through elastic and plastic deformation, and there is little pileup of atoms around the tip, as observed in the cross section. When the direction of the tool is reversed (lifted from the surface), the load quickly drops to zero, as expected. However, the load suddenly rises once again as the tool is removed further. This unexpected rise is caused by annealing at the surface of the metal so that the material returns to more intimate contact with the tool, and the load rises.

The results from simulations using silver as the indented material are quite similar to those just described for copper. However, because silver is softer, the yielding occurs at a smaller load than for copper.

Cutting is a widely used process in fabricating components. To obtain a more basic understanding of what happens when a metal is cut, a simplified geometry is used. We idealize the model slightly by limiting the motion to two dimensions or thin, periodic slabs in three dimensions. This idealization, known as orthogonal cutting, assumes that the material being cut is sufficiently wide that edge effects have essentially no influence on the results. Orthognal cutting is stimulated in both two and three dimensions. The 2D simulations more accurately model the length scales and time scales that are common in laboratory experiments using single point, diamond-turning machines. In this type of simulation, cutting speeds range from about 10 to 100 m/s. We have also varied the sharpness of the rigid tool tip used for cutting. We have used tool-edge radii of curvature ranging from 1 to 20 nm. The bluntest tip (20 nm) approaches the sharpest real-life tool tip, which has a radius of curvature of about 35 nm. A single frame from a computer-animated movie of the three-dimensional MD simulation of orthogonal cutting. The copper material flows from left to right at 100 m/ s. In this simulation, the carbon atoms in the diamond-like tool are not permitted to move, so they serve as the fixed frame of reference.

This type of projection makes it easier to see some important effects. During our simulation of cutting, the system forms a chip. The chip is reminiscent of what is actually observed during real experiments. The chip remains crystalline, but it has an orientation different from that of the surface. Regions of disorder among the atoms are apparent in front of the tool tip and on the surface in front of the chip. Unlike the 3D studies of point indentation, dislocations readily form during orthogonal cutting. Unlike the single point in the indentation simulations, the cutting tool in the orthogonal cutting geometry forms an infinite line, like a knife edge. This geometry provides much more energy for the creation of dislocations.

The calculations show that the cutting force strongly depends on the sharpness of the tool. This relation is also observed experimentally. Dull tools (those with a large tip radius) require larger forces to achieve the same depth of cut as sharp tools. A good measure to quantify the observations is the work (force x distance) performed by the cutting tool divided by the volume of material removed, also known as the specific energy. This energy increases dramatically with decreasing depths of cut. For our shallowest (nanometer scale) cut, the specific energy exceeds the energy required to vaporize the material, though the material remains a solid. The dependence of specific work on depth of cut has been observed in macroscopic metal cutting for many years and is known as the size effect. However, the macroscopic size effect is much less dramatic than the size effect observed in the simulations or in experiments using single-point, diamond-turning machines. The transition between macroscopic and microscopic behavior occurs at a length scale of a few micrometers, comparable to the average grain size in most metals. This result is interpreted as a change in the mechanisms of deformation, from grain-boundary sliding and motion of existing dislocations at the macroscopic scale to the creation of new dislocations and other point mechanisms of deformation at the microscopic, atomic scale. These dislocation-creation and point mechanisms of deformation consume significantly greater energies, leading to the observed size effect. In contrast to copper, materials like silicon and iron are not considered machinable with diamond tools. The reason is that the diamond tool wears rapidly, and it becomes difficult to maintain contour accuracy. The increased wear occurs for all materials that form strong chemical bonds (covalent bonds) with carbon. We have investigated and identified some of the underlying processes that give rise to these effects.

In covalent materials like silicon, the nature of interatomic forces is much different from that in metals. The strength of a covalent bond depends on the local environment, and the material forms an open structure. The basic method is the same as that for cutting copper, except that now we evolve not only the silicon surface material but also the carbon atoms in the diamond tool to allow for tool wear. It has been found that both copper and silicon show ductile behavior during cutting; however, the underlying mechanisms that allow this behavior are very different for the two materials. Whereas a copper chip remains crystalline, a silicon chip is transformed into a completely different state.

This time of a silicon-cutting simulation. In this case, the diamond wedge is cutting at a speed of 540 m/s. Although no wear of the tool is apparent yet, a single layer of silicon atoms has coated the diamond tip. Chip formation occurs between this layer of silicon atoms and the crystalline surface. In related simulations of diamond asperities (abrasive tips) scraping a silicon surface (as in grinding processes), it was found that the diamonds wear by forming small clusters of silicon carbide, which remain on the silicon surface. The silicon material in the chip and in the first few layers of newly cut surface appears to be amorphous or possibly to have melted. The temperature in the chip, calculated from the vibrational motion of silicon atoms, is comparable to the melting temperature of bulk silicon. However, the silicon atoms in the chip are not diffusing, which demonstrates that the material is, indeed, in a solid state. The following idea has been developed to explain chip formation in crystalline silicon. Because of the nature of covalent bonds in crystalline silicon, the energy needed to shear the crystal is enormous. In fact, less energy is required to transform the crystal into an amorphous solid and to shear the amorphous solid than to shear the crystal. The interplay between the energetics of different deformation processes and the dependence on length scale may explain the transition from brittle to ductile behavior that is observed in ceramic materials.

The ability to precisely machine ceramic surfaces like silica glass affects many different fabrication technologies and is central to constructing a wide range of sophisticated optics systems. A laser program at a University needs to make very high-quality mirrors, and we need to know more about special materials, such as ceramics, used in modern devices and equipment. We are applying MD methods to examine deformation processes at the atomic scale in these materials. The main difficulty is that glass and ceramic materials are brittle. Anyone who has seen a rock impact a windshield or a pane of window glass knows about the property of brittleness. This property can result in cracking, subsurface damage, and other kinds of costly damage during precision machining. However, a growing body of experimental evidence suggests that glass and ceramics can exhibit both ductile and brittle behavior during grinding, depending on the size of the abrasive used. In particular, there is an abrupt change in the surface smoothness of glass with a smaller abrasive size. In other words, with a smaller depth of cut, even glass can behave in a ductile manner. If we could better understand and manipulate this ductile-to-brittle transition, we could improve the economics associated with fabricating ceramic components. To address this problem, simulation of fused silica glass with no impurities was undertaken. In silica, four oxygen atoms surround each silicon atom to form a tetrahedron. Each oxygen atom is shared with two silicon atoms to form connected tetrahedra. Amorphous silica glass is a random network of these interconnected tetrahedra. To study the nanometerscale deformation of fused silica, a diamond tip was pushed with a radius of about one nanometer into a smooth silica surface.

It was found that the silica surface responds much more elastically than did either the copper or silicon surfaces. For indentations up to 1.25 nm, it was observed that no plastic deformation occured. At an indentation of 1.25 nm, a significant rearrangement of the silica network occurs directly beneath the tool tip. This rearrangement, an example of which is shown, leads to a decrease in the slope of the loading curve. When the tool is removed, the unloading curve does not follow the loading curve, as it does for elastic indentations up to 1.25 nm. The area between the two curves is a measure of the work performed by the tool in rearranging the silica network. These calculated loading and unloading curves are strikingly similar to those observed experimentally. It is essential to learn more about cracks for a better understanding of deformation in glass. To simulate the propagation of a crack into a silica surface, we put a notch on the top surface of the glass and let the computer simulation pull apart the sample at a constant velocity. The critical stress for crack propagation depends on the geometry and length of a pre-existing crack. Longer cracks, for example, yield at lower applied stress than do shorter cracks.

MEMS *and* Nano*technology Clearinghouse*

An information resource for the MEMS and Nanotechnology development community hosted by the MEMS and Nanotechnology Exchange, the nation's leading provider of high-quality foundry and consulting services

- News
- Events
- Jobs
- Resumes
- MEMS-talk

Controling the size of the particles is important because size is strongly linked to properties like electronic structure and melting temperature.

Arrays of silicon nanowires with biomolecular coatings can spot molecular traces of cancer.

Creating access to the most comprehensive array of foundry processes and software tools .

The market for tools used to fabricate emerging nanotechnology could grow from less than $20 million in 2004 to nearly $235 million by 2010.

Researchers at Dartmouth College have invented a robot so small that an entire army of 200 of them could march across an M&M .

Researchers exploring the potential of making and testing springs, rods, and beams at the nanoscale.

Collaboration is critical for getting nano solutions into clinics.

Joint Project Produces Unique Bacterial Diagnosis Application (2005)

Lab-on-a-chip application allows a more accurate detection of infectious diseases.

2005-10-07

- Scientists Learn To Prevent Nano 'merging': Controling the size of the particles is important because size is strongly linked to properties like electronic structure and melting temperature

2005-09-30

- Nano World: Nanowires Help Spot Cancer: Arrays of silicon nanowires with biomolecular coatings can spot molecular traces of cancer
- Coventor & Mems Exchange: Mems Tools Initiative Partnership: Creating access to the most comprehensive array of foundry processes and software tools

2005-09-21

- Nano World: Nano-tool Markets Rising: The market for tools used to fabricate emerging nanotechnology could grow from less than $20 million in 2004 to nearly $235 million by 2010

2005-09-16

- Tiny Robot Imitates Caterpillar: Researchers at Dartmouth College have invented a robot so small that an entire army of 200 of them could march across an M&M

2005-09-15

- Rensselaer Researchers To Study Nano Springs, Rods, Beams: Researchers exploring the potential of making and testing springs, rods, and beams at the nanoscale
- Living With Cancer: Collaboration is critical for getting nano solutions into clinics
- Joint Project Produces Unique Bacterial Diagnosis Application: Lab-on-a-chip application allows a more accurate detection of infectious diseases

2005-09-14

- Like Fireflies And Pendulum Clocks, Nano-oscillators Synchronize Their Behavior: Nanoscale oscillators could replace much bulkier and expensive components in microwave circuits
- Nano World: Public Attitudes Toward Nano: Public has low trust in government and industry regarding the health risks of nanotechnology

2005-09-09

- Nano World: Nano For Artificial Kidneys: Nano-filter mimics the function of the human kidney

2005-09-06

- Nano Shuttles Suggest Lifting Things May Become Thing Of The Past: Microlitre drops moved up a one millimetre 12 degree slope against the force of gravity

2005-09-01

- Nanospheres Block Pain Of Sensitive Teeth: Nanospheres could help dentists fill the tiny holes in our teeth that make them incredibly sensitive and that cause severe pain
- Mems Accelerometers Pick Up Speed: New sensors can discern more-complex and subtle three-dimensional movements
- Ce Growth Supercharging Mems Market: Pushing this growth are recent refinements in MEMS processing

2005-08-31

- Argonne Researchers Create New Diamond-nanotube Composite Material: New process "grows" diamond and carbon nanotubes together

2005-08-30

- Gold Bowties May Shed Light On Molecules And Other Nano-sized Objects: "Bowtie nanoantenna" greatly improves the optical mismatch between nanoscale objects and light
- Nano-tech Boffins Haven't The Foggiest: Polymer coating made of silica nanoparticles that can create clear glass surfaces that never become foggy
- New Microfluidics Device Can Deliver Targeted Anti-cancer Drugs To Brain Tumors: MEMS device is a candidate to deliver multiple drugs into the tumor microenvironment

2005-08-29

- Virginia Tech Researcher Reports Nano-particle Dispersion Technique Improves Polymers: Supercritical fluid carbon dioxide used; melt properties provide monitor
- Physicists Pose 'chip Dip' Nano Process: Nanotubes are deposited by dipping a chip covered with a gluelike substance into the nanotube solution

2005-08-24

- Nano-pen Writes In Tiny Letters: sharpened tip the size of a single atom applies an electric charge through a thin film of oil molecules onto a target surface

- Nano Diamonds Serve As Circuitry-writing Pens: Low wear and writing capability of ultra nanocrystalline tips with chemical inks is very promising

2005-08-22

- Nano Steel Will Light Up Batteries And Fuel Cells: Researchers have made an advance in nanotechnology by producing transparent carbon nanotube sheets that are stronger than the same-weight steel sheets

2005-08-19

- Nano Coalition Unveils Environmental, Health And Safety Database: Database marks the first effort to integrate the vast and diverse scientific literature on the impacts of nanoparticles

2005-08-17

- Nano-metrology To Probe Chip Structures At Atomic Level: Engineers are investigating a nanoscale approach to metrology that will allow them to examine new semiconductor structures at the atomic level

2005-08-16

- Nano-switches Could Yield Even Smaller Gadgets: Tiny Y-shaped tubes of carbon that act like electrical switches

2005-08-11

- Diamonds Are A Scientist's Best Friend: Tribological properties of diamond far exceed those of silicon
- Nano-sized Bomb Targets Tumors: Drug-delivering nanocell used to pinpoint and obliterate disease without harming healthy cells

2005-08-10

- Photons Are Exciting For Mems Disk: A micro-mechanical structure that vibrates when pumped with light has been developed at the California Institute of Technology

2005-08-09

- Nano World: A Semiconductor Nanotools Boom: Semiconductor industry tools and instruments that work on the nanoscale could form a $5.5 billion market by 2012
- Mems Work Uses Polysilicon-germanium Over Cmos: Gyroscope demonstrates combining a MEMS sensor and a signal conditioning circuit

2005-08-07

- Missed Opportunities In Nano: Nanotechnology companies are missing opportunities to help corporate buyers integrate nanoscale components into advanced products

2005-08-04

- Study May Expand Applied Benefits Of Super-hard Ceramics: Discovery could speed the design of materials that approach the hardness of diamond yet remain supple enough to be worked like metal

2005-08-03

- Magnetic Nano-oscillator Addresses Limitations Of Integrated Oscillators In Wireless Devices: Project aims at demonstrating the concept of spin torque in a nano-scale microwave integrated oscillator

2005-08-02

- Stress Testing At The Nano Scale: Material stress testing tool incorporates a flexure-based piezoelectric positioner, a capacitive probe displacement sensor, and a high-resolution load cell for testing of thin films under stress

2005-08-01

- Nano Silver Fights Infections: Silver nanoparticles could help fight hospital-related infections that afflict 2 million patients and lead to 90,000 deaths in the United States each year
- Micro, Nano Wrestle With Similar Challenges: Micro and Nanotechnology are both facets of science that address matter, devices, and products on a scale where the phenomena that govern their performance are different from traditional science.

2005-07-28

- Dna-based Molecular Nano-wires: Wires use derivatives of DNA with a greater electronic potential than DNA itself

2005-07-26

- Nano Detector Fingers Pathogens: A portable nano detection tool could be used by processors to ensure food safety

2005-07-21

- Ornl Mirrors Powerful Tools For Studying Micro/nano-materials: Precision mirrors to focus X-rays and neutron beams could speed the path to new materials and perhaps help explain why computers, cell phones and satellites go on the blink.

2005-07-20

- Nano-surgeons Break The Atomic Bond: The ultimate in surgery has been carried out in a vibration-free bunker
- New Mems Design Smokes Out False Alarms: New generation of fire detectors could reduce the rate of false alarms in the cargo and baggage compartments of commercial airliners.
- Two Steps Forward, One Back?: The nanotechnology concept is as old as the constituents of the pottery glazes used by the ancient Romans and as new as some of the coatings and composites used by automakers.
- Startups Rebuild Amid Telecom's Rubble: MEMS and nanotech startups re-emerge with overhauled products and business plans

2005-07-19

- Nanomemory: Commercial Opportunities for Nano-Based Memory and Storage Technologies

2005-07-15

- Ucla Chemists Create Nano Valve: UCLA chemists have created the first nano valve that can be opened and closed at will to trap and release molecules.
- Nano-graphite May Store H2 Gas: Graphite films only nanometers or billionths of a meter thick could help store hydrogen in an inexpensive, easily manufactured, lightweight and nontoxic manner

2005-07-06

- U. S. Risks Losing Nano Lead: The U. S. has fallen behind Asian competitors when spending levels are corrected to reflect exchange rates
- Nano-imprint Makes Its Mark: Imprint lithography used to operating circuits with a width of 30 nanometers

2005-07-01

- Nano World: Wiring Up Single Molecules: Carving of infinitesimal gaps into nanowires soon could help scientists connect electronics to single molecules

2005-06-23

- Buckyball Aggregates Are Soluble, Antibacterial: Research offers clues about C60 behavior in natural environments

2005-06-20

- Teleporting Over The Internet: Eventually, the teleported objects will be built with "nano-dust"—tiny objects that can be programmed to bind to each other and move

2005-06-17

- Nano Cluster Devices Unveils Hydrogen Sensor Prototype: Sensors could be used in a wide variety of applications

2005-06-13

- Nano World: Nano For Stem-cell Research: Nanotechnology can help regenerate organs with stem-cell biology and help people walk again and recover after strokes and heart attacks
- Brush Up On Your Nanotechnology: The world's smallest brushes have bristles more than a thousand times finer than a human hair

2005-06-12

- Nanotechnology Could Speed The Adoption Of Rfid Tags: Nanotechnology could really help accomplish the goal of five-cent RFIDs for ubiquitous use

2005-06-09

- Hp Reveals Design For Future Nano-electronic Circuits Packaging: Future nano-electronic circuits to use coding theory

2005-06-08

- Mit's Nanoprinter Could Mass-produce Nano-devices: Nature's most efficient printing technique: the DNA/RNA information transfer

2005-06-03

- New Self-assembling Technique Provides Path To Manufacturing Complex Nano-electronic Devices: Molecules arrange themselves to replicate the underlying pattern without imperfections

2005-06-02

- Interstate Effort To Boost Nano In Chesapeake Bay Region: Initiative will help make the Mid-Atlantic region a leader in the research, development and commercialization of nanotechnology

2005-06-01

- Magnetic Resonance Goes Nano: MRI device is small enough to fit on a computer chip
- Going Nano Boosts Thermoelectrics: Researchers use nanomaterials that allow only cold electrons to flow in order to minimize the amount of heat transfer

- Nano Leds Made Easier: Each lamp in the moth-eye array is 220 nanometers in diameter

2005-05-31

- Army's New Sensors Rock: The U.S. military is developing miniature electronic sensors disguised as rocks

2005-05-27

- The Future Of Nano-biology: Regenerating Tissue And Artificial Proteins: Every tissue from head to toe is being regenerated somewhere across the planet

2005-05-25

- Nasa Goes Nano For Air Purification: NASA is working to shrink the size of a device used to detect and measure pollutants in the air
- Nano-grating Dvds Could Store 100 Times More: reflective nano-structures to be used to encode data in a highly multi-level format

2005-05-23

- Nano Technology Tracks Cancer Spread: Microscopic new technology checks for the spread of cancer using magnetic resonance imaging machines found at most hospitals

2005-05-20

- Silicon Valve Holds Promise Of Fluid Controls Revolution: MEMS chip capable of controlling the refrigerant system in a car's air-conditioner

2005-05-19

- Nems Device Detects The Mass Of A Single Dna Molecule: Technology could be combined with microfluidics to perform genetic analysis of very small samples of DNA
- Report: U. S. On Right Path With Nano: The U.S. program to promote nanotechnology got a thumbs-up from its federal oversight body, which nevertheless said the NNI needs to take further steps to communicate and establish links to U.S. industry
- Quantum Dots: Nano-probes Of The Future: Scientists have created a procedure to render quantum dots water-soluble and non-toxic

2005-05-09

- Mems Startups Still Attractive, Says Vc Executive: MEMS design and manufacturing processes have yet to stabilize sufficiently to allow the

MEMS industry to divide into the fabless designers and the fabbed makers of components

- New Research Raises Questions About Buckyballs And The Environment: Buckyballs dissolve in water and could have a negative impact on soil bacteria
- Motorola Unveils Nano Emissive Flat Screen Technology: Optimized for large-screen high-definition televisions less than 1-inch thick, the prototype employs CNTs to change the design and fabrication of flat panel displays

2005-05-05

- Washington Kicks Off A Nano Initiative Of Another Stripe: Nanotechnology finally has something to offer the state's industries

2005-05-04

- Motor Transport In Bio-nano Systems: Micrometer bead is pulled by molecular motors, which are too small to be visible, along parallel filaments, which are immobilized on a substrate surface
- Nano Pyramids Boost Fuel Cells: One way to improve fuel cells that generate hydrogen on-the-fly is to increase the amount of surface area in a cell that can host the necessary chemical reactions

2005-05-02

- Nano World: Nano Could Hasten Hydrogen: Nanotechnology-assisted solar energy could help bring the dream of hydrogen-fueled vehicles to reality more quickly and cheaply

2005-04-28

- Researchers Devise Nano-scale Method For Investigating Living Systems: By observing how tiny specks of crystal move through the layers of a biological membrane, a team of electrical and computer engineers and biologists has devised a new method for investigating living systems on the molecular level

2005-04-27

- Nano Investors Facing 'implosion': Early investors in nano-technology may get their fingers burned

2005-04-20

- Nano Won't Ko Silicon, Says Moore: Nanotechnology will not replace silicon in the near future, according to Gordon Moore

2005-04-19

- Nano Startups Skip Across Chasm: If early 2005 is any guide, nano funding is headed up

2005-04-18

- Adaptive Optics Industry Finds New Markets Through Mems: Each mirror in micro-mirror array can independently tip, tilt and change position to smooth out aberrations in an optical beam
- Water, Water Everywhere: Nano Filtration: One of the single biggest applications of nanotechnology could be solving the global shortage of pure water

2005-04-15

- Wall Street: When Nano Firms Are Ready They'll Move Quickly: If Nanotech ever takes on Wall Street in a big way, it will appear to happen suddenly.
- Invasion Of The Poverty-fighting Nano-bots: Nanotechnologies that purify drinking water, produce energy and grow food can benefit poor countries

2005-04-13

- Mems Advances From A Bit-role Player To A Star Performer At Consumer Electronics Show: What a difference a year makes
- Tiny Engines Could One Day Be Used To Repair Damaged Cells: Relaxation oscillator is now the world's smallest motor

2005-04-12

- Nanotechnology's Miniature Answers To Developing World's Biggest Problems: Lab-on-a-chip could transform the lives of billions of the world's most vulnerable inhabitants

2005-04-08

- Tire Pressure: Ruling Seems Clear, But Read The Fine Print: A new tire pressure ruling appears to favor MEMS but over the long term could accommodate other technologies.
- Nano World: Nano For Quantum Computers: Possibilities for a nanotech quantum computer include: quivering nanotubes, superconducting nanocircuits and quantum dots

2005-04-07

- Ask The Analyst: Four micro/nano experts show what life's like inside market research

- Nano-agents That Strip For Action: New nanotechnology improves fuel efficiency

2005-04-04

- A Nano Drug's Giant Promise: After a decade-long slog to FDA approval, APP's Abraxane, a novel, less-toxic cancer treatment, has doctors and investors hopeful

2005-03-31

- Nano-probes Stay Inside A Cell's Nucleus For Days: Research could help biologists better understand DNA replication, genomic alterations, and cell cycle control

2005-03-29

- Toshiba's 'nanobattery' Recharges In Only One Minute: Battery fuses Toshiba's latest advances in nano-material technology for the electric devices sector with cumulative know-how in manufacturing lithium-ion battery cells

2005-03-28

- Mems Makers' Efforts Bear Fruit As Apple Embraces Sensor: Laptop computers will come equipped with a sudden motion sensor designed to prevent damage to sensitive components
- Nano-based Chips Won't Easily Replace Silicon: ...according to former Intel co-founder Moore

2005-03-18

- Nano Hazards: Exposure To Minute Particles Harms Lungs, Circulatory System: Several dozen reports unveiled detail how nanopollutants interact with the body
- Nano-probes Allow Inside Look At Cell Nuclei: Technique could be used to determine whether a drug has arrived where it is supposed to, and if it is having the desired impact
- Nano World: Water, Water Everywhere Nano: One of the single biggest applications of nanotechnology could be solving the global shortage of pure water
- Bacteria Bridges New Nano-circuit: Scientists in Wisconsin have used bacteria to make tiny bio-electronic circuits, a step towards new sensors and perhaps new nano-manufacturing techniques

2005-03-16

- Researchers Study How To Make Nanomaterial Industry Environmentally Sustainable: Study is the first to show what factors affect the size of aggregate particles
- Data Storage May Enter New Nanotech Phase: Itsy bitsy rods and disks may be able to store vast numbers of data bits
- Hp Plots Its Nano Course: HP is looking at nanotechnology to solve its future computing needs

2005-03-14

- In Solution, Tiny Magnetic Wires Scatter Light: Researchers believe "Nanowires" will one day become critical components in ever-shrinking electronic circuits

2005-03-12

- 'millipede' Small Scale Mems Prototype Shown At Cebit: Technology capable of achieving data storage densities of more than one terabit (1000 gigabit) per square inch

2005-03-10

- Mems Motor Means No More Dead Batteries: Tiny engines that generate power may one-day replace batteries
- Nano Particles Could Solve Ground Water Problem: Using nanotechnology, researchers were able to maximize the number of palladium atoms that come in contact with TCE molecules and improve efficiency by several orders of magnitude over bulk palladium catalysts

2005-03-08

- Sampling 'small Atmospheres' In The Tiny New Worlds Of Mems: Sandia gas sampling device rapidly determines whether MEMS seals are effective
- Nanobiotech: Nano Poised For Liftoff: First 'nano' technologies yield fruit in the lab and clinic with the promise of more to come

2005-03-07

- Investing In Nano: Venture Capitalists Tell Their Side Of The Story: Mark Brandt, Jeff Fagnan, Matthew McCall, Waqar Qureshi answer questions posed by Jeff Karoub

2005-03-06

- Early Cancer Detection Is Treatment Of The Future, Nano-medicine: The metallic nanoparticles are made especially for medical applications in a patented process

2005-03-04

- Nano World: Nano To Speed Drug Discovery: Nanotechnology could help more candidate drugs make it to market through clinical trials

2005-03-02

- Tracking In Locations Where Gps Won't Go: To improve accuracy, MEMS gyroscope uses three sensors: one on each leg and a third worn at the waist

2005-02-28

- Nano 'ruler' Could Provide Microchip Benchmark: A nanoscopic measuring device that uses atomic lattices to gauge tiny distances could enable microelectronics engineers to build better components

2005-02-25

- Nano World: Edible Nanotech On The Horizon: Edible forms of nanotechnology could help make smart programmable drinks and more effective drugs
- Nano-enabled Drug Discovery Set To Dominate: Nanotechnology has been singled out as a tool, which could make the difference in an industry

2005-02-23

- The Worries Over Nano No-nos: Could the same properties that make the tiny particles so effective also turn them into efficient troublemakers inside the human body?

2005-02-22

- Tiny Data Switch May Store More: Mechanical device is a completely new approach to improving data storage

2005-02-21

- Successful Test of Single Molecule Switch Opens the Door to Biomolecular Electronics: Developments due to approach based on measurements made with the molecules suspended in solution
- Startup Uses Tiny Probes To Store Data: Technology could take the $7 billion flash memory market in an entirely new direction

2005-02-18

- A Mason Finds His Muse In Nano's Building Blocks: NanoTechno series consists of images of molecular structures and materials—quantum dots, nanowires, carbon nanotubes and thin films

2005-02-17

- Nano Startups Consolidate Ip Positions: Small tech patenting activities—of which nanotechnology is a part—have skyrocketed in recent years

2005-02-15

- Nano Mechanism To Control Protein May Lead To New Protein Engineering: UCLA scientists have created a mechanism at the nanoscale to externally control the function and action of a protein
- Outfit Greases Wheels Of Nano Commercialization: Particles have a unique structure of nested spheres that lubricate by acting as miniscule roller balls

2005-02-14

- Nanobusiness Alliance: Budget Likely To Give Nano A Trim: Group says more focus needs to be put on funding the gap between the research stage and the point where most venture capital firms are willing to provide support for bringing a product to the marketplace

2005-02-11

- Nano World: Nanotech May Cut Pharma Waste: Much of the waste the pharmaceutical industry generates could be cut dramatically by using nanotechnology reactors
- Engineers Develop Biowarfare Sensing Elements: Devices Permit Mass Production of Highly Sensitive and Stable Nerve-Gas Detectors
- Devising Nano Vision For An Optical Microscope: Phase-sensitive, scatter-field optical imaging could extend life of conventional light-based imaging methods

2005-02-09

- Mems Rf Front-end Filters Needed: MEMS-based solutions have quality factors that exceed 10,000, well beyond conventional ceramic filters
- Side-impact Regulations Could Drive More Demand For Mems Sensors: Airbags like the side and curtain ones are expected to provide enhanced protection in passenger vehicles
- Mems Industry Group Releases '05 Report: Group noted a critical lack of MEMS characterization tools

2005-02-08

- Nano-sized Probes Allow Researchers To See Tumors Through Flesh And Skin: Technology will be less costly and more accessible than MRI-based methods and free of the harmful side effects associated with radioactivity

2005-02-07

- Nano And Chips: Uneasy Ties: Semiconductor execs bridle at the suggestion that this new technology could unseat them. It is, however, a distinct possibility

2005-02-04

- The Business Of Nanotech: There's still plenty of hype, but nanotechnology is finally moving from the lab to the marketplace
- New York's Big Hopes For Nano: With fabs and labs sprouting around the capital in Albany, the state is shaping up as a power in the emerging technology

2005-02-03

- Nano Surfaces Could Slash Cost Of Solar Energy: Light-trapping technologies could reduce the thickness of semiconductor materials needed in solar panels
- Battery Uses 'nano Needles': Nanotechnology provides a unique mechanism to combine the chemicals and control the characteristics of the chemical reaction

2005-02-02

- '04 Nano Funding Report: Less Money: Dollars were down, but nano deals increased from 34 in 2003 to 45 in 2004

2005-02-01

- Channeling Nano's Future: Startups will plot multiple courses to public markets this year
- Nanometer Schmanometer: Keep your eye on MEMS

2005-01-31

- Phased-array Antennas Get Mems Relays: MEMS components could go into lower-cost phased-array antennas for collision avoidance and adaptive cruise control systems in automobiles
- Florida Professor Leads Drive For Nano-enhanced Armor: Ballistic shields will be ready for field testing in three to six months

2005-01-28

- Nano's Road To The Future: 5-year-old National Nanotechnology Initiative keeps U.S. efforts on course...for now
- Danish Researchers Design Virtual Nano-catalyst: Research offers new opportunities in the fields of renewable energy, pollution control and in the chemical industry
- Nano Paint Could Boost Antiterrorism, Rescue Efforts: New technology may be used to detect cancer in the first cells to become malignant

2005-01-25

- Lab-on-a-chip Finally Catching On In Mems Field: Key applications include life science research areas of genomics, pharmacogenomics, and proteomics

2005-01-21

- 2005 May Be A Momentous Year For Mems: Or maybe just a momentum-building year
- Lotus Plant Inspiring Mems Technology: Researchers at Ohio State University have been taking a look at the lotus plant in a quest to reduce friction

2005-01-18

- A Giant Leap For Nano: American Pharmaceutical Partners, Bristol Myers Squibb battle for market share forcasted to be $3 billion
- Mems Researchers Make Supersensitive Motion Detector: Sandia researchers created a MEMS sensor that can detect motions smaller than one nanometer
- Musclebots Crawling In Petri Plate: Device is said to be better than micromotors as it does not requires any external power

2005-01-14

- Study Finds Advantages To Iron Nanoparticles For Environmental Clean Up: Iron nanoparticles may be effective in cleaning up carbon tetrachloride in contaminated groundwater

2005-01-12

- Nano Gas Turbine Designed: Nanoscale gas turbine made from nanotubes

2005-01-11

- New Nano-plastic Can 'see' In The Dark: Semiconductor crystals were dispersed in everyday solvents similar to how particles in paint behave
- Pins & Vias: Mems Enter The Mainstream: Use of devices based on MEMS in electronic products is expected to grow

2005-01-10

- Regional Recap: States Fund, Flaunt Nano Competitiveness: Nanotech appears to be on lawmakers' minds as state governors and legislators set their agendas for 2005

2005-01-07

- 'nano-scissors' Laser Shows Precise Surgical Capability: Tool opens new frontier for biologists studying nerve regeneration
- Nano World: Molecular Electronics Rising: Molecular electronics will be capable of doing things that silicon cannot
- Nano-drug Bombs Target Tumors: Nanoscale capsules may one day deliver cancer drugs directly to tumors
- Nano-propellers Sent For A Spin: Metallic rods about 500 times smaller than the width of a human hair have been turned into tiny "propellers"
- Fund Manager Advises Balancing Risk: EVEN IN A MICRO-NANO PORTFOLIO

2005-01-02

- Breakthrough In 'nano' Tracking Of Cancer Hailed: Cancer successfully predicted in 90% of early experiments
- Ucsb Scientists Build Nanoscale 'jigsaw' Puzzles Made Of Rna: Nine different RNA fabrics were generated by tectosquares self-assembly

2004-12-29

- Chip-scale Magnetic Sensor Draws On Mini Clock Design: Magnetometer is about as tall as a grain of rice

2004-12-28

- Nano Firms Tie The Knot: The future of carbon nanotubes just got a little more exciting

2004-12-26

- Nano-lubricant Could Mean No More Oil Changes: Lubricant to be based on spherical inorganic nanoparticles

2004-12-22

- Simmons Remakes Bed With Nano-enhanced Fabric: Bed features a removable mattress top that can be washed to rid the mattress of germs

2004-12-17

- Nano World: Nanomaterials Buyers Beware: More than 200 companies worldwide sell nanomaterials. As a group, they have a poor track record

- Polymer Science Enables New Breakthroughs In It And Nano Medicine

2004-12-14

- Tight Twist Toughens Nano Fiber: Multiwall nanotubes contain several layers of successively larger tubes
- Tiny Tubes For The Big Picture: Nano-materials could be used in medical imaging

2004-12-13

- Physicists Discover Potential In Ferroelectric Nanodisks And Nanorods: New phenomenon could open door to nano-memory

2004-12-10

- Survey Shows Public Can Discern Nano's Benefits: Nano advances have reached the point at which potential is becoming reality

2004-12-06

- Spilling the beans on new technology: You can now laugh at any drink thrown in your face because of nanotechnology
- Software, tools key markets in china: Chinese government selects Coventor design tools software for its MEMS program

2004-12-02

- Uc-berkeley to build nano research center: Center to be built from $11.9 million grant from the National Science Foundation

2004-12-01

- Multiphysics analysis speed mems development: In the absence of a superphysicist, all members of the team can help solve these multiphysics problems

2004-11-30

- Flash market offers mems ray of hope: Companies have developed storage technologies that couple a grid of MEMS cantilevers with atomic force microscope tips

2004-11-29

- Polymer made in a nano test tube: Fullerene epoxide polymerizes to form a linear chain inside a carbon nanotube

2004-11-28

- Model prospects: accelrys announces nano consortium: Goal is to accelerate the development of software for designing nanomaterials and nanodevices

2004-11-25

- Infineon shrinks transistors with nanotubes: Researchers grew nanotubes measuring no more than 1.1 nanometers in diameter in a controlled process

2004-11-24

- Mini generator has enough power to run electronics: New microengines would be smaller, last 10 times longer than batteries

2004-11-23

- Gold nano anchors put nanowires in their place: Rows of horizontal zinc-oxide nanowires grown on a sapphire surface are held in place with gold nanoparticles

2004-11-22

- Siemens begins work on silicon-based mems devices for monitoring building climate: Prototypes will measure the CO2 and ammonia content in mouse cages

2004-11-19

- Nano fabric may make computers thinner: Researchers in Russia and England who claim they have discovered the world's first single-atom-thick fabric
- Mems makers tune in to digital tv market: Among the contenders is a MEMS-based technology pioneered by Texas Instruments known as DLP

2004-11-18

- Nanotech could put a new spin on sports: "It's all about controlling the physics of how the ball spins"

2004-11-15

- Tunable takes a simpler tack: Tunable lasers should replace all fixed-wavelength lasers in communications applications
- New technology breathes life into your food: Samsung and LG devising micro and nano technology to prolong the shelf life of frozen foods

2004-11-12

- EPA Backs Nanomaterial Safety Research: Activists Say $4 Million Is Too Little for Studies

2004-11-10

- Gold nano anchors put nanowires in their place: Researchers at NIST have demonstrated a technique for growing single-crystal nanowires in place

2004-11-08

- Twosomes Become Troublesome If One Side Goes Under: Partnership inherently is a risky proposition, given the high percentage of MEMS firms that fail.
- CMOS-compatible SiGe micromachine: Belgian researchers have developed a deposition sequence of poly-SiGe at CMOS-compatible temperatures

2004-11-05

- Nanotechnology may change energy industry: Technology can make coal a cleaner source of energy

2004-11-03

- Molecules form nano containers: Research shows the self-assembly of minuscule multicompartment structures.

2004-11-02

- Tiny tools carve glass: Researchers develop microscopic carving instruments

2004-10-28

- Nano-product sales to be $2.6 trillion by 2014: Sizing Nanotechnology's Value Chain
- Nanotech Group's Invitations Turned Down by Environmentalists: None of the three invited representatives of environmental groups has agreed to join the newly created International Council on Nanotechnology

2004-10-27

- Nano Central: A VC looks at how nanotechnology is putting new life into old economy industries

2004-10-25

- Nano-Fabric Reveals Unique Properties: Electrons can travel without any scattering over submicron distances, an important property for making very-fast-switching transistors
- Kerry May Pursue More Nano Funding; Bush Expected To Continue Support: John Kerry favors increased funding for science and technology initiatives
- Hitachi devises MEMS chip to quickly mix liquids at set ratios: Chip is capable of processing tens of milliliters of liquid per minute

2004-10-21

- Serial Entrepreneurs: True tales from industry veterans
- Nano-bullet for non-invasive treatment of cancers: The advantage of using smaller particles is that they can be inserted into any part of the human body and treat cancer cells in their infancy

2004-10-20

- Researchers refine nano self-assembly: Device organizes itself into predictable patterns

2004-10-19

- Stabilizing Mems System Lifts Hovering Vehicle's Chances: MEMS devices help the organic air vehicle remain stable as it hovers above surveillance targets

2004-10-18

- Dreaming of a Single-Chip Mobile Phone: Integration ultimately leads to lower manufacturing costs

2004-10-12

- Trying Nano On for Size: Don't be alarmed, but nanotechnology is in your pants
- Smart shirt that can call for help: Jacket is fitted with a sensor and a Bluetooth transmitter that communicates with a receiver

2004-10-11

- HP bets inkjets will open new markets: Inkjet printing is based on a manufacturing technology that is a close cousin to MEMS

2004-10-06

- Mems, Aka 'big Nano,' Seeks To Elevate Its Political Profile: Director of Berkeley Sensor and Actuator Center sees technical and political convergence for micro and nanotechnology

2004-10-05

- High-tech tweezers enable nano-assembly lines: Technique used to put things together at the microscopic or atomic level
- Virus forms nano template: Self-assembly mechanisms bring together charged membranes and oppositely charged polymers

2004-10-04

- SEMI Publishes Six New Technical Standards: New standards were developed by equipment suppliers, device manufacturers and other companies

- Firms Get Over $20 Million For Micro And Nano Products: Funding makes it easier for companies to take on otherwise fiscally risky projects
- Device has keen sense for mini-motions: Sensor claimed to be 1,000 times better than existing technology

2004-10-01

- For Sensors, MEMS the Word: Get started experimenting with a MEMS sensor!

2004-09-30

- Unifying Approaches Sought At Mems Conference: Standards and reliability discussed at MEMS Industry Group meeting

2004-09-28

- Nano proponents square off against specter of 'gray goo': Experts say Nano field will contribute $1 trillion to the U.S. economy by 2015

2004-09-27

- Buckyballs: Study Finds Toxicity, And A Cure: Researchers at Rice have demonstrated a way to reduce the toxicity of water-soluble buckyballs by a factor of ten million
- Samsung Sets Nano Benchmark: Congomerate earned an estimated $779 million in revenue since its 2003 nano launch

2004-09-20

- Radant Takes Lead In Race To Bring Mems Rf To Market: Firm has a $5-million government grant to verify its switches
- Light excites nano-antenna: Antenna captures visible light like radio antennas capture radiowaves
- Emerging Nanophotonics Markets: New Revenues from Processing Light with Nano Devices

2004-09-17

- Mems May Make Smart Homes More Affordable: Thanks to MEMS, the basic building blocks actually are there for bringing intelligence into housing for the masses

2004-09-13

- Cancer Institute Starts Nanotechnology Drive: Nanotechnology offers new ways to detect, diagnose and to treat cancer at its earliest stages and with minimal side effects

2004-09-08

- DARPA Contracts Lucent for Maskless Litho: Lucent to design, develop and demonstrate MEMS-based Spatial Light Modulators to enable maskless optical lithography
- Nano memory scheme handles defects: Researchers from Hongik University in Korea have devised a memory architecture designed for nanoscale crossbar electronics

2004-09-07

- Nanotech in Fashion: The Trend in New Fabrics: A new wave of nanotechnology allows clothes to resist spills and wrinkles, and wear longer between washings
- Molecular Imaging Focuses on Nano-level Life Sciences: Atomic force microscopes (AFMs) allow scientists to peer at the nano world

2004-09-03

- Bookham adds overload protection to 10Gbit/s receiver: Integrated MEMS Variable Optical Attenuator controls the power reaching the optical receiver

2004-09-02

- Coms 2004 Kicks Off With Upbeat Look At State Of Industry: Killer MEMS applications remain an elusive target, a thriving nano/MEMS community continues to grow

2004-08-31

- The Daintiest Dynamos: By harvesting energy from radioactive specks, nuclear microbatteries could power tomorrow's cellphones
- Mems Researchers Perfect Fabrication Of Atomic Clock: Researchers at NIST have developed the core technology for a MEMS-based atomic clock
- Cell phones could keep atomic time: Atomic clocks may be headed into cell phones, thanks to a breakthrough by federal researchers.

2004-08-30

- The Players and Pretenders of Nanotech: Investment opportunities in the nanotech world
- Environmental Fate of Nanoparticles: A pressing question remains as to the environmental impact of manufactured nano-sized materials

2004-08-26

- Four startups gunning for TI's microdisplay business: Engineers foresee significant growth for microdisplay-based TVs

2004-08-25

- Nanotech will tap nature's potential: Humans can learn a lot from animals, particularly the microscopic life forms living in thermal vents

2004-08-19

- Nano-coating cuts phthalate migration: Researchers say that the material cuts phthalate migration in medical applications.

2004-08-18

- Nano Discovery to Make Internet 1000 Times Faster: Canadian researchers have shown that nanotechnology can be used to pave the way to a supercharged Internet based entirely on light.
- Wireless Sensors: How Not to Replace 1000 Batteries: Wireless microsensors will be used to monitor everything from nesting conditions for endangered birds to stress-induced cracks in large structures. The big question? How to make batteries for them that last for years.

2004-08-05

- Nanosys Says No to IPO: On second thought, never mind.
- MEMS makes a nonvolatile-memory match: The MEMS memory works by flexing a microbeam.
- Embedding A Mobile Phone In Your Ear: If you thought people walking down the street talking into their bluetooth headsets looked odd enough, just wait until mobile phones are completely embedded in your ear.

2004-08-02

- Princeton Researchers Use Nanoimprinting for Mass Producing Nano Scale Devices: Low-cost production method may lead to greater memory capacity and lower costs for computers, digital cameras and other devices
- Science of the small could create 'nano-divide': New technologies could increase the divide between rich and poor nations

2004-07-30

- Federal funds go to national underdogs in nano research: Scientists and entrepreneurs working together to promote Oregon take note of $5 million in likely federal funding and a $1.3 million grant

- Nanotechnology Precaution Is Urged: Minuscule Particles in Cosmetics May Pose Health Risk, British Scientists Say

2004-07-29

- Myths and realities of nano futures: Ever since John Dalton proved the existence of atoms in 1803, scientists have wanted to do things with them

2004-07-27

- The promise and perils of the nanotech revolution: Possibilities range from disaster to advances in medicine, space
- Advancements in micro technology catapult MEMS-based applications: The hunt for the 'next big thing' is on

2004-07-21

- VIA to launch Nano-ITX embedded motherboards in 2H: Nano-ITX motherboards will have dimensions of 12×12cm
- Small Fuel Cells 2004, an event review: Small Fuel Cells 2004, the sixth in a series of annual symposia concentrating on portable applications, took place in Arlington VA, USA, in May 2004.

2004-07-19

- Green Goo: The New Nano-Threat: Nanobiotech carries the power to create completely new organisms that have never existed
- Three new standards for MEMS devices: Standards help researchers measure more accurately material characteristics used in construction

2004-07-16

- IBM claims nano-scale imaging breakthrough: 'Micro-cantilever' is 1,000 times thinner than a human hair
- USC Scientist Invents Technique To Grow Superconducting And Magnetic 'Nanocables': Nanocables manufactured from a potent new class of substances with extraordinary properties called Transition Metal Oxides
- Mems Update: 2004 Likely To Be Banner Year: The first half of 2004 is looking a lot like the last half of 2003 for the MEMS industry, only better

2004-07-12

- Konarka: Giant Leap With Nano-based Solar Power: Semiconducting particles of titanium dioxide coated with light-absorbing dyes used to convert light into electricity
- Active Spectrum Switches Gears With Wireless: Using MEMS technology, Active Spectrum wants to make a multi-frequency transceiver

2004-07-09

- A Nickel Investment For Future's Grid Will Pay Off: Energy is the single most important challenge facing humanity today
- Nano-team spins tomorrow's yarn: A method to continuously spin nano-material has been developed by Cambridge-MIT Institute scientists

2004-07-08

- University Develops 12Tbyte Nano Memory: Crystals are 10nm across and can be deposited in a single crystal layer on a substrate
- Is Telecom Ready To Embrace Mems-based Solutions?: Cautious optimism—the market is now finally beyond bottom

2004-07-06

- European Union Earmarks Billions To Prepare For A Nano Future: Technology is crucial to the development of mobile communications device makers

2004-07-01

- Oregon Blazes Nano Trail With More Money, Grants And Events: ONAMI in the running for $10 million in federal defense grants for 2005
- Tech Company Gets Hypersensitive: Sensor technology claimed 10,000 times more sensitive than what's available today

2004-06-29

- MEMS Industry Finding Success One Step at a Time Reports In-Stat/MDR: Venture capital funding in the first half of 2004 alone has already exceeded that provided in all of 2003
- Hubble's Early Demise Sends Nanotech Back To Waiting List: Cancelled mission to the Hubble Space Telescope would have been first applied nanotech in space

2004-06-17

- MEMS Enjoying Revenue Growth, VC Attention: 2003 revenues in the MEMS industry reached $5.3 billion, up 35.7 percent year over year
- MEMS-Driven Wearable Display For Auto Techs In The Works: Head-up display lets an automobile service technician consult reference manuals without stepping away from the vehicle
- Mems Sensors May Speed Work Of Jet Engine Designers: Sensors measure the fluctuating pressure as air passes through the inlets and mixes in the engine chambers.

2004-06-10

- Kodak's New Image: Century-old Firm Develops Nano Strategy: Company has developed pigment nanoparticles as small as 10 nanometers

2004-06-07

- IEEE and SEMI Sign Agreement to Foster Nanotechnology and MEMS Standards: The agreement marks the first standards collaboration between the two organizations.
- MEMS Come to Oz Wine Industry: Sensors are distributed across a vineyard and send their information by wireless link

2004-06-04

- Amateur Scientists Study Microscopy In Basement Nanolabs: Amateur scientists satisfy their curiosity through kitchen chemistry, backyard stargazing and garage physics

2004-06-03

- Too Little Too Late?: Report Gives Brits Failing Nano Grade

2004-05-28

- GI Joe Goes Nano: Soldier's items could be lightened, combined or integrated.
- Scientists use nano-sized particles: Particles one-billionth of a meter in size provide a "stable imaging marker" during surgery to remove brain tumors

2004-05-25

- Scaling Friction Down to the Nano/Micro Realm: Friction measurements made at the micro- and nano-scale can differ substantially due to changes in applied load.

2004-05-21

- Keystone cops: MEMS accelerometers outlaw projector flaw

2004-05-18

- The Way We Live: Nanotechnology and Medicine

2004-05-17

- Gyricon Is Xerox's Poster Child For Copy Company's New Nano Image: Etch-a-Sketch examplifies commercial small tech from the "The Document Company."
- Language of Science Lags Behind Nanotech: Burgeoning Field in Need of Universal Way to Describe Creations

2004-05-14

- Big Blue Says Breakthrough Means Millipede May Crawl Out Of Lab: The Millipede chip holds 4,096 miniature read-write heads

2004-05-11

- Intel Debuts 90 Nano Mobile Chip: The 90 nanometer process produces smaller, faster transistors

2004-05-10

- Chiral Photonics Brings A New Twist To Optical Devices, Lasers: 15,000 helical twists per inch in the core of the fiber areeach about 2 microns in width

2004-05-09

- Nano's Great Leap: A safer approach is to find the companies that sell "enabler" technologies

2004-05-07

- Nano-scale trees created at Lund Institute of Technology: Research demonstrates the possibility of producing more complex nano-scale structures

2004-05-03

- Postal Service Delays Rollout Of Cepheid's Anthrax Detector: GeneXpert testing system detects trace levels of DNA.

2004-04-30

- Nano Wires Make Tiny Compasses: Compass needles are just under half the size of the previously smallest known compasses
- Nano Weapons Join the Fight Against Cancer: "Nanoshells" and other ultrasmall tools could improve diagnosis and treatment of tumors

2004-04-29

- High-speed Nanotube Transistors Could Lead To Better Cell Phones, Faster Computers: Findings add to mounting enthusiasm about nanotechnology's revolutionary potential
- A Conveyor Belt for the Nano-Age: Someday, nanoscale conveyor belts could expedite the atom-by-atom construction of the world's smallest devices

2004-04-26

- Vermont's Seldon Labs Wants To Keep Soldiers' Water Pure: Firm to develop a water filter that uses carbon nanotubes

2004-04-19

- Living On A Bootstrap Budget, Luna Is Proud Parent To Five Spinoffs: Prototype products made into commercial development with outside partners
- China develops first nano-satellite: Super-small satellite weighs 25 kg

2004-04-16

- Immunicon's Nano IPO Gets 'sane' Response From Investors: Cytometer that enumerates and differentiates between the immuno-magnetically selected cells based on the fluorescence signature of the cells
- Technion Develops Nano-Sized Vehicles to Detect Gas Contamination: Device was modeled on dandelion seeds enabling them to fly in the wind

2004-04-13

- Nanomaterials Production Weaves Job Growth In Former Textile Town: 400,000-square-foot tobacco warehouse converted for production of nanomaterials

2004-04-12

- Nano Ribbons Coil into Rings: Tiny rings are potentially useful as sensors, resonators, transducers and actuators
- Nanoparticles Player Stirs The Pot With Room Temperature Process: ANW's process can yield 2-nanometer zinc oxide particles.

2004-04-09

- Environmental Journal Publishes Controversial Buckyball Study: Rates of brain damage to be 17 times higher in nine large-mouth bass

2004-04-07

- ?stentennas' Could Signal Change In How Arteries Are Monitored: A "stentenna" keeps arteries free of plaque and monitors blood flow and pressure.

2004-04-06

- Mems Technology Magnifies Opportunities For Low-cost Sem: NOVESEM uses MEMS to make less expensive components for a tabletop SEM.

2004-04-05

- Ancient Greeks help scientists build environmentally friendly nano devices: Technique inspired by the lost-wax casting process used by ancient Greeks for sculpture
- Mems Gyroscopes Making Strides In Replacing Entrenched Technologies:

MEMS gyroscopes quickly matching the performance parameters of older gyro technologies

- New Kyocera Company Shows Industry's Oled Intent: The LG4050 has an OLED subdisplay outside and a conventional screen inside.

2004-04-02

- Latest Gadgets And Future ?must Haves' Seen At Consumer Show: Texas Instruments remains leader with DLP technology for digital TV

2004-04-01

- 'Nano-Lightning' Could Be Harnessed To Cool Future Computers: "micro-scale ion-driven airflow" device uses a sort of nano-lightning to create tiny wind currents

2004-03-31

- Tiny machines need even tinier lubricants: Gases to provide microscale thin films of slippery coating.
- Mems Doesn't Spell Auto Opportunity, Despite Market Report: French report shows MEMS gyros and accelerometers market is growing

2004-03-29

- Millennial Net Hopes To Hook Up 21st Century Sensor Networks: Radio-connected sensors self-organize into wireless networks
- Nanoparticles Toxic in Aquatic Habitat, Study Finds: "nanoparticles" can trigger organ damage and other toxic effects

2004-03-26

- A Few Billion Years In Development, Biomimetics Emerge At Last: Velcro is the quintessential example of biomimetics

2004-03-25

- Businesses Begin To Pay Attention To Nanoparticle Health Debates: Nanotox raises questions about the toxicity of nanoparticles

2004-03-24

- 'Nano-lightning' could cool future computers: New technology creates tiny wind currents
- DNA has Nano Building in Hand: Researchers used DNA branch migration to construct the DNA hand.

2004-03-23

- Where's The Road To The 'last Mile'? Perhaps In The Packaging: Industry watchers predict 2004 will see progress in getting "fiber to the home"

2004-03-22

- Technology Conceived In The Cold War Changes With Times At Sionex: Sensor product can be calibrated to identify explosives, germs or other materials

2004-03-19

- Hey, Is That A Nanocapsule On Your Face?: Line of cosmetics contains nanocapsules, which help active ingredients get to the skin's deeper layers

2004-03-18

- Atomic microscope spots viruses: Nanotechnology used to develop a very small silicon chip to catch and identify viruses
- Hernia Patient Can Continue Active Life With Nano-enhanced Device: TiMESH, contains a 30-nanometer-thick coating on a standard plastic mesh implanted to repair damaged tissue

2004-03-15

- Small Times Names Top 10 Small Tech Hot Spots: Which states lead the race to become hot spots in nanotechnology, MEMS and microsystems?
- The Small and the Beautiful: Max Planck researchers use nanotechnology to visualize cellular processes crucial for the development of new cancer drugs

2004-03-12

- Effects of nanotechnology on the environment: There is still uncertainty about the impact of releasing nanoparticles into the environment

2004-03-11

- Little Things Could Mean a Lot: Nestlé is screaming for smoother ice cream...

2004-03-10

- Intel's research chief tells nano crowd to think even smaller: Think very small: 10 nanometers or less, says Intel's David Tennenhouse.

2004-03-09

- Nanolab Knows That Profits Are Not Made On Nanotubes Alone: And, yes, the company does make a profit.

2004-03-03

- Precisely Manipulating Millions of Atoms: Method may soon help scientists build nanoscale devices much more rapidly
- Sliding Without Friction: Friction can be controlled at the nanometer scale

2004-03-01

- One Startup, Two Employees And One Really Smart Membrane: Filaments can be customized to bind to specific targets

2004-02-25

- Profiting From Nanotechnology: Overview of what the science is and where opportunity may lie for investors

2004-02-24

- New Index To Track Nano Stocks, But Large-caps Stay Off For Now: Investors interested in nanotechnology finally get what they want: a stock index.

2004-02-23

- Nanomagnetics Has New Materials In Store For Memory Market: DataInk captures metal granules in a shell of ferritin protein with an 8-nanometer inner cavity.

2004-02-20

- Gyros To Go: SENSORS WILL KEEP MOVING FOR MILITARY, AUTO
- Quantum Cryptography Companies Tap Into Nanoscale's Quirky Core: Curious physics of individual photons, puts unique power of nanoscale to work.

2004-02-17

- 3D fabrication uses light-activated molecules to create complex microstructures: Technique could compete with existing processes for fabricating microfluidic devices

2004-02-16

- Water-based Nano-reagent for Molecular Biology Research Tool Kits: Neowater(TM) enhances performance of molecular biology tool kits
- STMicro Announces 3-Axis MEMS-based Accelerometer: Device provides both three-axis sensing in a single package and a digital output
- Personal medicine in waiting room: U.S. Genomics finds smaller market

2004-02-12

- Nano-origami: Scientists at Scripps research create single, clonable strand of DNA that folds into an octahedron
- Mems Usa Hires Executive, Going Public: Reverse merger is seen as a quick and efficient way of going public
- For Nanoart To Imitate Real Life, Exhibition Goes Back To Basics: Nano exhibit inspired by the "cell", capable of self-sustaining existence and replication.

2004-02-11

- Nanomaterials Could Soup Up Supercapacitors If Price Is Right: It's a battery! It's a power pump! It's Super Capacitor!

2004-02-10

- Stents rivalry to intensify as Boston Scientific enters market: Rival developers of drug-releasing stents have been battling for supremacy on the patent front. Now, there are issues of safety on the patient front as the two-way tussle moves to the marketplace.

2004-02-09

- Optical switches primed for test debut: DirectLight platform to target telecommunications, manufacturing, as well as fiber distribution, production and storage
- Nano Patterning: IBM brings closer to reality chips that put themselves together
- New Microscope Shows Nano-Fibre Formation: Microscopy probes nanofiber growth process with angstrom resolution

2004-02-06

- Industries await dawning of nanotechnology age: Stain-resistant fabrics and fresh food packaging only the beginning
- Incorporate Disassembly Into Every Self-assembled Nanotech Product: Molecular nanotechnology may clean up the toxic mess left by older technologies

2004-02-04

- Treatment Could Remove Toxins In Blood Before Damage Occurs: Nanoscale magnetic particles being developed for medical applications.

2004-02-03

- President Thinks A Billion Of Nano, But A Bit Less Than Allowed By Law: Proposed 2005 budget would give nano programs a bit less than the amount authorized under the bill.

2004-02-02

- Evolved Nanomaterial Sciences Separates The Bad From The Good: While most nanotech companies attempt to build things up from the bottom, Evolved Nanomaterial Sciences Inc. is approaching the nanoscale from the opposite end.
- Flexible Display Screens Readied for Production: Roll 'Em, Fold 'Em, Stick 'Em in Your Pocket: Long-Envisioned Plastic Sheets Will Make E-Newspapers, E-Books a Reality

2004-02-01

- For Science, Nanotech Poses Big Unknowns: Studies show nanoparticles can act as poisons in the environment and accumulate in animal organs.

2004-01-30

- Nanotennis Anyone? Tiny, Sporty Materials Have Their Day On Court: VS Nanotube racket is responsive when you are out of position and can barely get your frame on a ball.

2004-01-29

- Coatings Company Wants Consumers To Look For The U-right Label: NANOECO sportswear is treated with Texcote technology, which places a layer of protective coating on the surface.

2004-01-28

- Self-assembly Technique Shines Even If Flash Device Was For Show: Technique for making nanocrystals could become a cornerstone in IBM's future miniaturized components.

2004-01-27

- Quantum Control Holds The Key To A Shining LED Lighting Market: Market for high-brightness LEDs projected to grow from $1.53 to $2.85 billion

2004-01-22

- Asia-pacific Governments Invest In Nano Labs And Research Centers: Singapore's Biopolis research park will host a mix of public research institutes and private biomedical companies.

2004-01-20

- Molecular memory not a dream as ZettaCore attracts real money: Tip of the molecular electronics wedge may ultimately evolve into mainstream nanoelectronics

2004-01-19

- Pursuit Of A Small Tech Building Block Goes Down To The Nanowire: When is a wire more than a wire? When you add the "nano" prefix, of course.

2004-01-15

- British Scientist Says Nanoparticles Might Move From Mom To Fetus: Vyvyan Howard study could flag up a new, particular, hazard of nanoparticles

2004-01-13

- Small Tech Tackles Concussion Syndrome With Mems Helmets: MEMS accelerometers to create an "impact history" for each wearer.

2004-01-12

- Nanomed Pharmaceuticals' Co-founders Invade Inner Space: Nanotechnology is used to package drugs in tiny spheres and send them to specific cells

2004-01-08

- Europe's MEMS Research Program Surprisingly Spawns Startups: Licensed technologies create new businesses.

2004-01-07

- Nanosphere's newest detector zeros in on specific diseases: Latest gadget could someday detect AIDS and Alzheimer's disease.

2004-01-06

- Philips finds an e-paper technology that's quicker on the draw: Thin electronic-paper display uses a process called "electrowetting"

2003-12-30

- Hitachi set to plant its own 'nanostamp' on the medical market: Process uses a silicon "stamp" to produce long, thin "nanopillars."

2003-12-22

- Universal display wants to shine its OLED light on the world: Organic light-emitting device technology to see the light of day in 2004

2003-12-16

- Geeky garments keep you comfy, and just might save your life: Who says geeks can't look good?
- IntelliSense Software Launched: IntelliSense Software Announces Acquisition of Software Business Unit Assets of Corning IntelliSense

2003-12-15

- Here's a new spin (cycle) on small tech: Smart appliances: Over the next several years, "intelligent" appliances will hit home

2003-12-10

- Ibm's Directed Self-assembly One Facet In Broad Nanotech Program: Chuck Black and Kathryn Guarini demonstrated their nanotech breakthrough with strands of purple and red pipe-cleaner wire
- Nano Europe Goes Down To The Crossroads Between East And West: TRIESTE, Italy has become the temporary home for more than 1,000 nanophiles this week.

2003-12-08

- Microreactors could redefine chemistry, nanodrip by drop: Move over lab-on-a-chip, make room for chemical-plant-in-a-box.

2003-12-04

- EV Group and AMO GmbH to Cooperate on Nanoimprint Lithography: Agreement on low cost, high resolution, large area patterning process

2003-12-03

- IPD library from PHS MEMS supports Agilent's EDA software: New Integrated Passive Devices library designed for Agilent's electronic design automation software

2003-12-02

- Boob tubes in your Buick?: RF MEMS-based antenna arrays may make it possible
- New IBM ThinkPad retains data even if owner is thoughtless: MEMS accelerometers may soon protect data

2003-11-26

- MEM Research Releases EM3DS 6.2 Beta: Now including 3D and 2.5D engines

2003-11-20

- Small tech VC in Q3: The story remains staid and sustainable

2003-11-19

- Investors Take a Serious Look at Nanotech as Senate Passes Nanotechnology Bill: Investorideas.com features News, Research, Links, Company Profiles For Nanotechnology Stocks
- India selects Coventor MEMS software for national program: India's MEMS Initiative Program chooses CoventorWare? for new design centers

2003-11-17

- MEMS / MST magazine free of charge: by Yole Developpement
- DALSA selects EV Group as Strategic Supplier for MEMS Equipment: Wafer Bonding & Thick Polymer Lithography Production Equipment to be provided

2003-11-14

- DELFMEMS: the MEMS solution !: DELFMEMS provides services in Micro/Nanotechnologies.

2003-11-13

- PHS MEMS Integrated Passive Devices (IPD) confirm expectations: OEMs have recognized improved performance and testability by using this state of the art technology for several applications

2003-11-12

- Microfabrica to Enhance Microdevice Developments in Japanese Market: Secures strategic distribution alliances with Itochu Corporation and Sumitomo Corporation

2003-11-11

- Southeast France plants seeds for a fertile small tech future: France is beginning to look more like the French version of Silicon Valley

2003-11-10

- Small Times magazine recognizes small tech in today's world: Small Times announces award winners
- McGill University selects EV Group equipment to build nanotools: McGill University's Nanotools Facility will be formally inaugurated on November 10th, 2003

2003-11-07

- Nanotechnology educational materials: available for free online

2003-11-04

- Gold nano bullets shoot down cancer tumours: Tests on human breast cancers, both in the test tube and in tumours in mice, were highly successful

2003-11-03

- Play It Smart: Sensor makers create 'cool' and cut costs to pry open dog-eat-dog toy market
- MEMS industry recognizes packaging can make or break a product: The technical difficulties involved in packaging can be considerable
- Partnering makes difference for packaging, testing, assembly
- Company's IPO shows good form, but it will be a tough act to follow
- Bantamweights vs. giants: Is lab-on-a-chip brawl a mismatch?
- Molybdenum products and others: Molybdenum products and others
- Silicon Carbide heating element: SiC heating elements
- Molybdenum disilicide heating elements: MoSi2 heating elements

2003-10-31

- EV Group and SiGen forge wafer bonding alliance on Next-Generation Plasma Bonding

2003-10-30

- FLX Micro Secures Additional Venture Financing: November 3, 2003
- Tronic's Microsystems And Sercel Present The Results of Their Collaboration In Mems-based Seismic Sensor Production: Tronic's industrial and technological expertise enables a new generation of geophones for land-based oil and gas exploration
- Tronic's Microsystems Opens New Mems Manufacturing Facility
- polyimide film from China: insulating plastic films

2003-10-29

- EV Group's aligner and bonder enhance biochip research at ASU

2003-10-26

- Integrated Micro Devices Corporation provides rapid polymer MEMS prototyping

2003-10-22

- Indian fab to pursue smart-card, MEMS manufacturing

2003-10-20

- NEW SUSS nano PREP TECHNOLOGY FIRST TO REVOLUTIONIZE DIRECT WAFER BONDING
- IMEC and EV Group will jointly develop new wafer-level packaging and MEMS wafer bonding techniques

2003-10-14

- IMT Welcomes Dr. Ian Johnston From STS

2003-10-13

- Coventor To Assist Intellisense Customers With Software Upgrade: CoventorWare? conversion package provides MEMS focused training and support
- China Selects Coventor As Preferred Vendor For Government Mems Development Program: CoventorWare software to be supplied to 30 universities through China's MEMS/863 program

2003-10-10

- MEMtronics Receives $3.69 Million DARPA Contract for RF MEMS Switch Development
- Numbers May Not Gauge It, But MEMS Industry Is Moving At Midyear

2003-10-09

- Block MEMS Receives $2 Million Government Contract to Develop Miniaturized Chemical Agent Detector (ChemPen) for Battlefield and Homeland Defense

2003-10-07

- RM in Life Science Applications: Breakthrough in micro fluidic devcies

2003-09-30

- MICRO SYSTEM Technologies 2003 Best Paper and Best Poster Awards: MICRO SYSTEM Technologies 2003 Conference and Exhibition, October 7-8, Munich Germany

2003-09-28

- MoSi2 heating element and SiC heating elements: MoSi2 heating element and SiC heating elements

2003-09-26

- DISCERA Selects DALSA for Manufacture of Quartz Replacement Technology: Relationship Leverages Low-Cost, Low-Power Benefits of RF MEMS
- Silicon Microstructures (SMI) announces miniature pressure sensor chip
- Silicon Microstructures (SMI) announces new pressure sensor chip with improved media compatibility
- Silicon Microstructures (SMI) releases first in its family of co-integrated sensors
- Fraunhofer Institute Selects Substrate Bonder from SUSS to Support MEMS Commercialization Program

2003-09-17

- SUSS MicroTec's New Substrate Bonders Offer Superior Process Control: Bonders enable high yields and ensure functionality of MEMS, compound semiconductors

2003-09-16

- Flow Science Announces Release Of Flow-3d® Version 8.2
- Silicon Microstructures (SMI) completes expansion and renovation of its MEMS wafer fab: Previously announced multi-million dollar plan completed 2 months ahead of schedule
- MEMS-based Explosive Detector Could Catch Future 'shoe Bombers'
- Micronit Microfluidics Signed Agreement On High Volume Lab-on-chip Foundry Manufacturing

2003-09-06

- FLX Micro Enters Exclusive Licensing Agreement With Case Western Reserve University

2003-09-03

- Memgen changes name to Microfabrica

2003-08-26

- MEMS Standards, While Small, May Mean Much For The Industry
- Molybenum Products: Molybdenum Powder, Molybdenum electrodes...
- MoSi2 heating elements and others

2003-08-25

- Axsun Isn't Sitting By The Phone, Waiting For Telecom To Call Again

2003-08-22

- PhoeniX invests in MST software: PhoeniX develops software for Micro System Technology to provide the bridge between the design team and the real cleanroom environment

2003-08-20

- Tire Sensor Makers Gauging Pressures Of Court Decision

2003-08-18

- FLX Micro Names Bob Lynch As Chief Executive Officer: August 20, 2003

2003-08-12

- Thick Negative Photoresist which can be removed: SU8 Alternative

2003-08-11

- Flow Science Announces New Associate
- Palomar Hopes To Dominate Assembly With Its One-stop Shop

2003-08-08

- EV Group presents new company website www.EVGroup.com

2003-08-04

- STS Sharpens Its MEMS Tools To Carve A Niche In Industry

2003-08-01

- Not Exactly Quicker Than A Ray Of Light, But Optical MEMS Will Fly

2003-07-29

- Integrated Micromachines Equipment Auction

2003-07-28

- atomic IC and nanocomposites: atomic electromechanical (AEM) systems, nanocomposites
- Jmar's In Motion To Grab A Corner Of The Microchips Bag

2003-07-21

- Anti-stiction layer deposition in a vapor phase: samples solicitation

2003-07-16

- Flow Science Announces New Associate
- Dutch MEMS: Strong Research, But No Mass Production . . . Yet

2003-07-15

- Advanced Microsystem for RF and Millimeter wave Communications: AMICOM "Network of Excellence" Within 6th Framework

2003-07-14

- With New Cash, Facility And Focus, Tronic's Tries For New Markets

2003-07-09

- EV Group Sells Production Lithography Package To Silex

2003-07-08

- A Delicious Idea — But It May Be Too Pricey To Be Palatable

2003-07-07

- Amkor Provides Glimpse Of Future IC-packages
- Hybrid MEMS 3D Dynamics Analyser
- Cea-leti Selects EV Group's Hot Embossing System To Conduct Nanotechnology Research

2003-07-01

- New Purpose Built 12" Mems Foundry In Nickel: Photo Electroformed Nickel MST

2003-06-30

- Olympus' Partnership Across The Pacific Expands Its Mems Borders

2003-06-23

- Agency Launches Full-service MEMS Practice

2003-06-18

- Micralyne, Canada Collaborate To Mass-produce Mems On The Cheap

2003-06-17

- Silex Microsystems has attracted USD 9 million

2003-06-16

- Nexense:: Direct Digital Measurement

2003-06-13

- Coal-Mine Canaries on a Chip
- Ultrasound Works Like Bat's Ears To Make Real-time 3-d Images

2003-06-12

- MEMGen Unveils '3-D MEMS Design Challenge' Winners: Top three prizes include free prototypes and up to $10,000 cash

2003-06-11

- Researchers create potential toxic sensor chip by combining electronics with living cell

2003-06-10

- EV Group introduces a new wafer edge inspection system at SEMICON West

2003-06-09

- STS positions platform for MEMS volume etch
- SiSonic surface mount microphone to appear in mobile devices soon
- RF MEMS tools and IP to grow from European project
- Immense potential lies in amazingly small things

2003-06-06

- XACTIX to Unveil Xetch eXplorerTM: Will make breakthrough MEMS release process obtainable for a wider range of programs

2003-06-05

- Hitachi Enters A House Haunted With Ghosts Of Optical Past

2003-06-04

- Discera Begins Sampling Wireless Industry's First Micro-oscillator: First Single-Die Multi-Frequency Device Will Replace Multiple Quartz Crystals

2003-06-03

- Discera Joins Center for Wireless Integrated Microsystems (WIMS): Microcomponent Developer Strengthens Ties to Innovative University Group
- Discera Joins Berkeley Sensor & Actuator Center: BSAC Affiliation Provides Access to Latest Research Results and Technology Commercialization Opportunities

2003-05-29

- Software Firms Keep Hard Times At Bay With New Business Models

2003-05-23

- DARPA funds "smart-dust" research
- Onstream Bankrupt Again, Leaving Mems Projects Homeless

2003-05-22

- Neonode Multimedia Phone To Use Emkay Sisonic MEMS Mike

2003-05-20

- Innovative Micro Technology Welcomes Dr. Amir R. Mirza from GE NovaSensor
- EV Group's temporary bonding and debonding technology enables thinned wafer processing in high volume manufacturing

2003-05-19

- Read/Write Heads Teach Advanced Microsensors Small Tech 'rithmatic

2003-05-17

- MEMS/NEMS suppliers database available on the website of Intelligent MEMS Design, Inc.

2003-05-16

- LioniX moves to new premises
- LioniX Forms Alliance with Indes: Companies to develop and integrate microsystems

2003-05-14

- PHS MEMS Launches Custom Integrated Passive Devices For Wireless Applications: Design Services and Volume Manufacturing Process Enable Miniaturization and Cost Reduction within RF Modules
- The Latest Developments In Microsystems, MEMS, MOEMS And Microfluidics At Micro System Technologies 2003 In Munich: Conference program now decided
- Xerox Runs This Up The (Telephone) Pole: MEMS For Magical 'Last Mile'

2003-05-07

- TRONIC's Microsystems unveils new MEMS production facility
- Small tech battle-tested in Iraq in prelude to warfare's new wave

2003-05-05

- Abtech Scientific uses small tech to sense the future of medicine

2003-05-01

- What can Europe teach America? a little self-help, for starters

2003-04-30

- FLX Micro issues "call for designs" for its third MUSiC run

2003-04-29

- Small tech companies outduel all comers at Purdue competition

2003-04-28

- Its cells may be thin, but Cymbet says they pack a big punch
- MEM Research announces FBAR modeling capabilities: Ver. 6.1 of EM3DS includes full-wave modeling of Bulk Acoustic Wave Resonators
- ALCATEL signs exclusive agreement with CANON SALES on Japanese MEMS market

2003-04-25

- As privacy vs. security debate heats up, NSF primes sensor pump

2003-04-23

- Discera Speeds to Commercialization of Wireless Microcomponents with New Hires and California Office: Product and Manufacturing Managers To Help Launch Miniaturized Low-Power Products

2003-04-22

- Here's a new idea: the Internet; customers demand e-commerce

2003-04-21

- FlyTimer jettisons half of 'a wing and a prayer' for small planes

2003-04-17

- IMMI introduces a new low cost fluxless wafer-level MEMS packaging

2003-04-15

- MEMGen Extends Deadline for MEMS Design Challenge to April 30, 2003

2003-04-11

- Officially, war kills nano summit; but some suspect it went AWOL

2003-04-10

- OAI Installs New Back-side Mask Aligner at TFI: Automated, versatile aligner facilitates manufacturing

2003-04-09

- Motors of light may resolve dilemma of how to power MEMS
- Developing motors of light to power small electronic devices
- Memsic finalizes plan for China facility

2003-04-07

- IMMI is switching from telecom to more promising products

2003-04-02

- Small Times Media names Crosby as new president and publisher
- Dutch company positions itself for the growing MEMS market

2003-04-01

- STMicroelectronics puts some marketing muscle into BioMEMS

2003-03-27

- Emkay's SiSonic Microphone Designed in Neonode's State-of-the-Art Mobile Phone

2003-03-25

- Deadline Nears for MEMGen's MEMS Design Challenge: Prizes include free prototypes and up to $10,000 cash
- MEM Research releases EM3DS 6.0: The Free limited and the full trial versions are available for download

2003-03-24

- EnOcean Proffers Piezo Power so Low-energy Products see the Light
- EV Group supports Waseda University on Nanotechnology Research

2003-03-19

- Coventor DESIGNER Trial License Program Announced: Layout and modeling software gives design teams MEMS capability
- CoventorWare 2003 MEMS Design Training Announced: Course Fee Waived through July 2003
- New Motown MEMS Lab Will Steer Delphi Beyond the Automobile

2003-03-18

- WSU unveils MEMS lab

2003-03-17

- Small Times Magazine Names Top 10 Small Tech Hot Spots
- Nanoledge builds on nanotube knowledge to make macro material
- LNL mans the photonic torpedoes to prepare for telecom's return

2003-03-13

- U.S. nanotech funding expected to hit $1 billion
- Good things, small packages: Binghamton University electronics engineering center charts new directions at 'micro' and 'nano' scale

2003-03-12

- California, Massachusetts Locked In Bicoastal Clash for Supremacy

2003-03-07

- MEMS Worn to be Wild, But Air Bag Vests a Hard Sell to Easy Riders

2003-03-06

- OMM Dies In Telecom's Waiting Room; Customers Mourn While IP's On Ice

2003-03-05

- Hannover Fair 2003: Micro-Scale High-Tech for Competitive Products: 37 International Exhibitors on the IVAM-Pavilion Present Product and System Solutions

2003-03-04

- MEMSCAP and DOLPHIN partner to bring high level language solutions for MEMS simulation to market: MEMSMaster deploys MEMSCAP modeling expertise and powerful design environment enabled by Dolphin's SMASH single engine, mixed-signal, multi-level and multi-language simulator.

2003-03-03

- Flow Science Announces New Associate
- Bioprocessors Breaks a Bottleneck In the Drug-discovery Process
- EV Group Celebrates Opening of New Joint Venture Company in Taiwan

2003-02-27

- New Opportunities for Optical MEMS Are Emerging Outside of Telecom, Says New CIR Report

2003-02-26

- Not Content to Gather Dust In the Lab, Pioneer Brings Motes to Market

2003-02-25

- Nanotech to pave way for micro-machines

2003-02-20

- Nanoimprint Equipment Companies Set to Make Market Impression

2003-02-19

- Intel donates 150mm wafer line to New Mexico MEMS project

2003-02-17

- EVG highlights 300 mm Automated IQ Aligner at SEMICON Europa 2003

2003-02-14

- New MEMS Startup Offers Design, Modeling and Fabrication Services as Well as MEMS Equipment and Patent Services: MEMS and ASIC service, MEMS equipment, IP service
- After Columbia: Small Tech Can Help Make Space Travel Safer

2003-02-12

- New MEMS Startup Offers Design, Modeling and Fabrication Services
- MEMS to remain a niche technology?
- Kionix Partners with Kefico to Supply MEMS Accelerometers
- SoCs are "dead," Intel manager declares
- Survey Finds "disconnects" Between MEMS Developers and Fabs
- EV Group ships volume MEMS production equipment to Asia Pacific Microsystems Inc.

2003-02-11

- Calient Networks to provide Optical Switches for "OptIPuter"
- Coventor Teams up with Cadence
- Agilent Invests in MEMX

2003-02-10

- Inventor throws curve at screens
- Corning IntelliSense, Northrop Grumman team on RF MEMS

2003-02-04

- TRONIC's Microsystems Appoints Peter Pfluger as CEO: Former CEO Stephane Renard Takes on New Role as Executive Chairman
- Small Tech's Voice Heard Amid Clamor of Competing Agendas

2003-01-31

- Brewer Science increases focus on MEMS market
- Emkay Releases SiSonic—the World's First Surface Mount Microphone: Based on MEMS Semiconductor Technology
- EV Group strengthens its activities in the advanced packaging market.

2003-01-30

- Patent Promises to Place Smart Scalpels in Surgeons' Hands

2003-01-29

- In Telecom's Long, Dark Winter, MEMS Survivors Keep it Simple

2003-01-23

- MEMS/Nano LIGA Division Launched by Digital Matrix

2003-01-22

- Small Tech Auto Applications Begin Their Trip in High-end Vehicles

2003-01-20

- Silicon Microstructures (SMI) embarks on major MEMS facility expansion: Relocation of Los Angeles etch operations to Milpitas facility; Multi-million dollar expansion of Milpitas MEMS facility to increase capacity and add capability
- Miniature Tool & Die Is Helping to Mold the Micro Market

2003-01-17

- EV Group provides industry first 300mm volume production wafer bonding system

2003-01-16

- CoventorWare 2003 MEMS Design Software Available

2003-01-15

- MEMGen Ignites Micro-Device Innovation: Kicks off MEMS Design Challenge; Prizes include free prototypes and up to $10,000 cash
- Massive Data Storage in Tiny Devices

- MEMS meet semiconductors in beginning of a beautiful friendship

2003-01-14

- New Penn State Summer Institute for Biomaterials and Bionanotechnology: A summer research opportunity for undergraduates and graduate students

2003-01-10

- MEMS industry hopes to hire, while veterans see growing pains

2003-01-09

- Companies put their MEMS together to drum up more customers

2003-01-08

- How to start a nanotechnology (or MEMS) business

2003-01-06

- Iolon's MEMS-Driven Lasers Reduce Need for Spare Telecom 'Tires'

2003-01-03

- Tiny radio signal switches set to turn on RF MEMS market

2002-12-30

- PamGene says it's found the fast track to gene research products

2002-12-27

- European study says only the big photonics players will survive

2002-12-23

- Microlab seeks funding as it prepares launch of RF MEMS switch

2002-12-20

- Companies Race to Develop Ultimate Detection Technology in Terror War

2002-12-19

- Swedish MEMS Fab SILEX MICROSYSTEMS installs ALCATEL AMS 200 ICP Silicon Etcher: Annecy, France (19th December 2002).
- Belgian center spins off company to sell drug-screening technology

2002-12-18

- MUSiC Call for Designs

2002-12-13

- Tanner Research Announces Availability Of MEMS R&D and Prototyping Services

2002-12-12

- MUMPs Update from MEMSCAP: New worldwide academic pricing, design rules, run schedules available for MUMPs

2002-12-11

- Innovative Micro Technology wins $1.8m DARPA Award for Developing MEMS Cell Purifier
- SUSS MicroTec introduces Embossing Equipment for Nanoimprinting Technology

2002-12-10

- UK Govt. Requests Proposals for Spectrum Efficiency Scheme
- Small tech gets a larger piece of the U.S. defense research pie

2002-12-05

- Inventor's long, strange trip to market fueled by MEMS
- Japanese Wafer Manufacturers Produce with SOI Wafer Production Bonding Systems from EVG

2002-12-03

- Alcatel introduces AMS 200 DE: New Deep Oxide Etching ICP Production Tool for MEMS & MOEMS

2002-11-26

- Startups selected for DARPA RF MEMS program

2002-11-22

- Swiss MEMS fab navigates new funding while it guides missiles

2002-11-21

- For MEMS, the key to lowering cost is all in the packaging

2002-11-20

- Trikon Intros Deep Silicon Etch Module for MEMS Devices
- FLX Micro Advances the World's First Micromachining Process in Silicon Carbide
- OnStream MST and Coventor Announce MEMS Partnership: RF MEMS Volume Manufacturing Ramp to Begin in 2003

2002-11-19

- New MEMS Market Research Report: MEMS Technology: Where To?

- MEMS really is the word at Munich electronics trade show

2002-11-18

- Cepheid poised for postal contract, plows on with life science plans

2002-11-15

- Sandia National Laboratories MEMS/LIGA Research Selects Digital Matrix

2002-11-13

- OMM to Supply Optical Switching Modules to ZTE of China
- Small Times magazine announces Best of Small Tech Award winners
- Few firms have figured out how to make MEMS quickly and cheaply

2002-11-12

- IVAM: Positive Signs for Hannover Fair: Industrial Solutions ?en miniature" are much sought-after

2002-11-11

- STMicro shows dual-function DNA analysis chip
- Coventor co-invents products while it reinvents itself

2002-11-10

- MEMS Etcher: An economical XeF2 etching system

2002-11-07

- EV Group Annouces New EVG®40 Series Top-to-Bottom Side Alignment Accuracy Measurement Systems

2002-11-05

- MEMSCAP Finalizes JDSU/Cronos Acquisition and Fab Financing
- Radioactive isotopes fuel microscopic battery
- Thomson Derwent Launches MEMS Patents Profile: New Resource Targeted At Engineers, Scientists and Researchers Working in Micro Devices

2002-11-04

- MEMS Supplier Silicon Microstructures (SMI) hires Mike Dunbar for Strategic Marketing
- Telecom is next on the continuum for 'smart ceramic' MEMS company

2002-11-01

- STMicro prototypes DNA analysis chip

- Display start-up projects new image
- Next Generation Data Storage on the Nanometer-Scale

2002-10-31

- Photonicsourcing launches a powerful search engine for photonic related products

2002-10-29

- MEMS give farmers the straight story on productive planting

2002-10-28

- Corning IntelliSense and Technovena Pte. Ltd. Partner to Develop Chinese CAD for MEMS Market: Market distribution partnership to strengthen support for Chinese customers of IntelliSuite®

2002-10-25

- 'MEMS' the word for new STAR Center project: Research aims to build miniature instruments

2002-10-24

- OAI Installs Back-side Aligner: Automated, versatile aligner facilitates MEMS manufacturing

2002-10-22

- Thousand-chamber biochip debuts
- Breast cancer diagnostic tool attracts interest, but needs cash

2002-10-21

- Support for EUV lithography grows
- Flow Science Announces Release of FLOW-3D® Version 8.1

2002-10-18

- Akustica Raises $2.25 Million in Series A Funding: MEMS Technology Company Spun-Off from Carnegie Mellon University
- Tiny atomic battery developed at Cornell: Could run for decades unattended, powering sensors or machines

2002-10-16

- Corning IntelliSense Corporation Awarded $2 Million from NIST Advanced Technology Program.: Research to advance development of digitally controlled MEMS.

2002-10-15

- Generics forms alliances with Coventor, Epigem and LioniX: Launching a new microsystems development service, MST-charged(tm).
- With prototype, funding in hand, Arradial ready to discover drugs

2002-10-14

- Chips & Drivers: MEMS are making memorable strides toward ubiquity
- Teravicta teams with Dow-Key on RF MEMS switches
- EVG highlights Production Resist Processing System for Spin Coating and Developing at SEMICON Japan 2002

2002-10-10

- University of Florida researchers make progress on tiny battery
- Expansion of Silicon Microstructures, Inc. MEMS Fabrication facility and equipment.
- U.S. announces technology funding recipients

2002-10-09

- MEMS event calendar now available in iCalendar format: Usable from Apple's iCal, Mozilla calendar.

2002-10-08

- "Terahertz" camera prototype developed
- Veeco Came, Saw, Acquired Majority Of The AFM Market

2002-10-03

- SEMI's Oasis provides respite from GDSII

2002-10-02

- TRONIC's Microsystems Announces Agreements
- IMT Partners with L-3 Communications for MEMS-based Inertial Measurement Unit
- For Triad, it's not the MEMS, it's the Method of Making Them

2002-10-01

- Cadence buys IBM's design-for-test tools business
- Analog Devices spins a monolithic MEMS gyro

2002-09-30

- MEMS chipmaker makes $1.9M in series C funding

- MEMS poised to move into the mainstream
- Expanding MEMS with low-cost SOI wafers
- RF MEMS Market to Exceed US$ 1 Billion by 2007: Market study by WTC

2002-09-27

- CFDStudio/ANSWER Version 5.5 is released

2002-09-26

- Smart Artillery Shells Promise a Major MEMS Device Market

2002-09-25

- MEMS Foundry Service Prospers In Southern California

2002-09-24

- XinTec Inc. Expands Shellcase Chip Scale Packaging Technology with EV Group's Production Mask Aligner and Resist Coater

2002-09-20

- Philips puts small tech to the test: It needs to beat other technologies

2002-09-17

- Newsweek on New Technology Jobs: "Fields of Dreams: New Job Opportunities"

2002-09-11

- Chicago area small tech firms are getting their act together
- Military calling all MEMS as U.S. beats Iraq war drums
- Small tech provides some answers to a nation questioning its safety

2002-09-10

- Michigan will help lead the drive toward small tech, governor says

2002-09-09

- Silicon Light Machines Introduces Reconfigurable Blocking Filter: Wavelength-Agile Optical Module Boosts Flexibility and Reduces Costs for Service Providers
- Michigan is host to COMS 2002, gathering of microsystems leaders
- Warning: Loose lips sink IP; shut up and patent, say experts

2002-09-06

- New CVD technique produces thick, high-quality diamond films

2002-09-05

- MEMSIC Debuts ADI pin-for-pin Compatible Accelerometer: New temperature compensated accelerometers offer 50% cost savings and industry-best 50,000g shock survivability rating
- New technology to seal MEMS could speed integration with chips
- Where have all the brainiacs gone? Ph.D. dearth may stymie small tech

2002-09-04

- Coventor Launches Teragenics: Spin-off Will Combine Distinctive Microfluidics Technology With Biotech Intellectual Property

2002-08-30

- Nanotechnology creates ideal optical medium
- Wireless MEMS are loud and clear while telecom suffers static

2002-08-29

- OLEDs get ready to light up the market for flexible screens

2002-08-27

- Kodak tries to sneak up on TI with its digital cinema strategy

2002-08-26

- CALMEC hopes to build a company a single-molecule switch at a time

2002-08-23

- Team has many obstacles to leap before the Doctor-on-a-Chip is in

2002-08-15

- Agere to dump optical business, leaving MEMS in state of limbo

2002-08-14

- Rise and Decline of Standard MEMS: Was It Just 'Too Much, Too Early'?

2002-08-13

- MEMS Micromirrors Reflecting the Big Picture in Home Theater

2002-08-12

- IMT Receives Investment from L-3 Communications for MEMS: A Strategic Partnership Formed
- Flow Science Announces New Associate
- Harris & Harris Nixes 'Tiny' Transformation

- Proteomics and Promises, Promises: Drug Makers Not Easily Convinced
- What Can Small Tech Do for Its Country? Just Ask, Says Top Soldier

2002-08-01

- MEMS Exchange: Number of Enlisted Fab Sites Keeps Climbing
- MEMSCAP TO ACQUIRE CRONOS MEMS UNIT FROM JDS UNIPHASE: Company To Become Exclusive Supplier of MEMS-Based Products to JDS Uniphase

2002-07-31

- N.J. Governor Funds Small Tech Consortium Based at Bell Labs
- High-speed Data Transfer with Microsystems Technology: IVAM Workshop Micro-optics/Opto-electronics at Dortmund

2002-07-30

- Coventor Announces Enhanced Electrostatic Solver: MemCap Turbo Module Now Shipping
- Swiss School Keeps Spinning Out Small Tech Firms at Dizzying Pace

2002-07-29

- Small Tech Funding Pitches Sought for COMS 2002 Conference

2002-07-26

- Amersham to Expand Microarrays after Buying System from Motorola
- EV Group installs wafer bonding system at Analog Devices: EV Group's SOI wafer bonding system provides volume, quality and yield enhancements at Analog Devices' SOI wafer production fab

2002-07-25

- India Ready to Launch Its First Privately Funded MEMS Lab

2002-07-24

- Firm Finds New Polymer Process

2002-07-22

- PHS MEMS Counts on Growth in Market for Mobile Handsets

2002-07-16

- Cronos Introduces Two New Multi-User MEMS Processes: MUMPs Program Expands to Include SOI and MetalMUMPs

2002-07-15

- Partech International Investment to Boost MEMGen Growth: Leads Series A1 financing

2002-07-12

- Gaming may be the 'Trojan Horse' that sneaks MEMS into your Palm
- Microstrain's orientation sensor gets the attention of the U.S. Navy

2002-07-10

- Teravicta Technologies to Offer RF MEMS-Based Products: Company's RF MEMS Switch Currently in Evaluation at 12 Leading Wireless Companies

2002-07-01

- Coventor Demonstrates MEMS Switches
- EV Group showcases Automated Spray and Spin Coating System at Semicon West

2002-06-10

- PowerZyme says it has the juice to play in portable power market
- MicroMD opens as part of a technology initiative

2002-06-05

- MEMSIC Awarded Best of Show at Sensors Expo: Ultra Low Noise sensor recognized for innovation by leading sensor publication

2002-06-04

- Small Times Launches Small Tech Census: The Most Comprehensive Measurement of the Small Tech Industry
- TRONIC'S Microsystems SA and the CEA have signed a MEMS R&D agreement
- U.S. military's research agency gets a new microsystems chief: Clark Nguyen joins DARPA
- BigBangwidth sets out to create a new universe of smart switches

2002-05-30

- Micralyne Reports Large MEMS Order from Automotive Sector
- APLAC Introduces MEMS Library

2002-05-29

- Micralyne may be a model for making money in MEMS

2002-05-28

- Brits give firm license to share

2002-05-24

- MEMSCAP Strengthens its Position in Asia: Long Term Deal with BL Marketing Services

2002-05-21

- Laser Micro-machining for MEMS and MEOMS

2002-05-20

- MEMS focused issue of Materials Today magazine: Materials Today for tomorrow's technology

2002-05-16

- MEMSIC Accelerometers Selected By Master Brake: Leading brake systems supplier cuts costs and improves performance of lightweight trailers using MEMSIC dual axis accelerometers
- MEMSIC Delivers Accelerometer With Lowest Noise Floor: MEMSIC only vendor to integrate circuit and sensor on one chip based on standard CMOS process to lower cost and maximize reliability

2002-05-07

- Critical Point Dryer for MEMS: Gentle and economic method for drying MEMS
- New Tight Genes Look Good As Better Disease Treatments
- Genefluidics Counts on Glass to Break Into Nanobio Market
- Biodegradable Nanofiber Could Prevent Scar Tissue
- Time For Physical Sciences To Get Its Act Together, Says D.C. Counsel
- French Venture Capitalists Keen on Nano, But Pinch Their Euros
- Electronic Nose Sniffs Out New Markets for California Firm
- Plasma Process for Particles Could Fuel Explosive Growth

2002-04-26

- Mr. Micromachine: Before They Called It MEMS, There Was Kurtz

2002-04-22

- Key to Life Science Success Is Market, Not Technology

2002-04-19

- Better Living Through Small Tech: MEMS and Nanotechnology Meet Real Life
- New Advanced Deep Plasma Etching Tool for High Volume Production of MEMS & MOEMS: Alcatel introduces AMS 200 "I-Speeder" at Semicon Europa

2002-04-17

- Solutions Come in Small Packages for German Firm with New Process

2002-04-16

- Hannover Fair 2002: Microtechnology Road Maps Point In Different Directions

2002-04-15

- Report from Hannover Fair 2002: Microtech Section is Small, But It's 'Where the Action Is'
- Lion Photonix and 3T mst group join their efforts: Lion Photonix Technologies BV and the business-unit Micro System Technologies of 3T BV, both based in Enschede, The Netherlands have joined forces for the development of Lab-on-a-chip systems.

2002-04-09

- Sound Holding changes its name to Sonion

2002-04-08

- New Microsystems Applications Drive Fast Growth, Study Says

2002-04-05

- Flow Science Announces New Associate
- MEM Research Releases Ver 5.2 of EM3DS: EM3DS 5.2 available for free download. Also available a free MEMS Gallery

2002-04-04

- Firms Lay Financial Foundation Resting on Nanoclay Composites

2002-04-02

- MEMS Fabrication Services: RF and Optical MEMS Manufacturing

2002-04-01

- Three-part series on higher ed programs in small tech

2002-03-26

- Relax, Your Car's Sensors Soon Will Be Better Than Your Senses
- German Nano, Bio Join Together To Try Building A Better Biochip

2002-03-21

- You've Got MEMS Under Your Skin? Sensor to Track Your Every Move
- New Lasers Tuned With MEMS May Make Nets More Efficient
- Program Tries to Kick-Start European Microtech Industry

2002-03-19

- 1x10 MEMS Mirror Array Available from Coventor and Colibrys

2002-03-14

- Small Times Analysis: Silicon Valley Could Rise Again as the Small Tech Hub

2002-03-13

- FiberLead Launches High Precision Optical Fiber Arrays

2002-03-12

- What Telecom Slowdown? MEMS Maker Debuts Product
- MEMS Robots to Give Surgeons Sense of Touch, Long-Distance

2002-03-11

- Flow Science, Inc. Announces New Associate
- Nanotech Reality Check: New Report Tries to Cut Hype, Keep Numbers Real: Nanotechnology Opportunity Report released today

2002-03-09

- microTEC presents New Solutions For Packaging And Interconnection Technologies

2002-03-08

- U.S. Regulators Want to Know Whether Nanotech Can Pollute

2002-03-07

- Micralyne Enhances Microfluidics Consumables Offering with Launch of ProtoChip

2002-03-06

- Noted Photonics Conference to Honor Nobel Laureates

2002-02-28

- For MEMS Only: An Evening at the Micromachines Club

2002-02-27

- Investors Put Faith, Money Into Microlab's MEMS Switch

2002-02-25

- LLNL Develops Powerful New Rechargable Battery
- Handheld Disease Detector Could Be Developed in 3 Years

2002-02-22

- Dispatch from the Telecom Shakeout: Auction Sells Remnants of MEMS Firm

2002-02-21

- French System Ensures Good Research Translates Into Marketable Products

2002-02-20

- Microvision Piles Up Patents in Retinal Display Technology

2002-02-19

- Swiss MEMS Company Sees Profitability in Flexibility: Colibrys profile; Corning Intellisense also highlighted

2002-02-18

- Altair Hopes for Lift from Pigment Patent

2002-02-15

- U.S. Law on Tire Sensors Skids Between Agencies

2002-02-14

- SphereTeK Solder Microball Technology: New low cost solder microballs for $3.00 per wafer

2002-02-13

- IMT Expands Fab and Capabilities
- Innovative Micro Technology Appoints New VP of Marketing and Sales
- Making MEMS: The Working Life of Kionix's New Microdevice Fab

2002-02-04

- Applied MEMS Teams with Kistler Instruments: Low Noise Accelerometer to be Offered

2002-02-01

- Umachines introduces Digital 1x2 MEMS Switch: Umachines is currently sampling this 1x2 characterized by low insertion loss, low power consumption, low PDL and lifetimes greater than 100 million cycles.

2002-01-29

- Flow Science, Inc. Announces Release Of Version 8.0 With Multi-Block Functionality
- Analog Devices Selects Coventor to Integrate Design Tools into Optical iMEMS® Development Environment: New MEMS development tools will reduce ADI's optical iMEMS® product design cycles

2002-01-23

- Micro Industry Involved in Power Struggle; Small, Cheap Energy Will Emerge Victorious

2002-01-22

- Record Numbers at MEMS Conference 'Means the Industry Is Really Forming'

2002-01-21

- Coventor Introduces Low-Cost Variable Optical Attenuator Integration Alternative
- MEMS Market Continues to Grow, Says Industry Group's New Report

2002-01-17

- MEMS Optical Names Angelo V. Ugge as CEO

2002-01-15

- Lion Photonix and OnStream MST combine their efforts

2002-01-14

- New structuring (USHD) and Polishing method (MDF) for glass and fused silica wafers

2002-01-10

- MEMSCAP Still Thinks Large Despite Telecom's Troubles: MEMSCAP building world's biggest MEMS factory

2002-01-09

- Coventor and SensFab Team for MEMS Product Design and Manufacturing: Relationship Gives MEMS Integrators Access to Proven MEMS Production Capability

2001-12-27

- MEMSCAP ACQUIRES CAPTO: New MEMSCAP Unit Brings Medical and Aerospace Business

2001-12-26

- First Book on Microfluidics: New Textbook/monograph

2001-12-17

- BTG and Coventor to Commercialize MEMS IP

2001-12-10

- MEM Research Releases EM3DS 5.0x: Trial version available for download

2001-12-07

- Alcatel Micro Machining Systems expands in Japan: New application laboratory in Tokyo

2001-12-05

- PHS MEMS Chooses Coventor ARCHITECT Tool for MEMS Development: Selection Key to Reducing Time to Market for Optical and Wireless Products

2001-12-04

- Teen Inventors Earn Scholarships by Making MEMS Devices Run Better: Siemens Westinghouse Science and Technology Competition Winners Announced

2001-12-03

- Primaxx Single Wafer MEMS Etch Release System: Vapor phase "no stiction" MEMS release

2001-11-29

- Introduction to MEMS Short Course: Offered by Carnegie Mellon University; Pittsburgh, PA

2001-11-28

- Beauty is Skin-deep with High-Tech Sensors: MEMSCAP and La Licorne Enter Multi-Million Dollar Partnership

2001-11-26

- MEMS Optical, Inc. Announces the Completion of a Major Expansion
- MEMS Collaboration Accelerates Time to Market: Agreement between Colibrys and Coventor will yield evaluation samples.

2001-11-12

- Kionix, Inc. Announces Completion of MEMS Fabrication Facility: Press Release 12 Nov, 2001

2001-11-10

- Microstructures by "click": Online Quotations

2001-10-17

- LIGA Now Available Through MEMS Exchange

2001-10-15

- Coventor Named One of Fastest Growing Companies by Inc Magazine: Coventor is First MEMS Company To Receive the Award
- NJIT Aiding Development Of DNA Microarray Technology

2001-10-13

- 1 Micron Precision Automation System

2001-10-12

- Micro-Map 5000: New interferometer available

2001-10-10

- MEMS: Mechanics and Measurements Call for Papers: Event Date: June 10-12, 2002 Abstracts Due: October 2001

2001-10-05

- Alcatel opens new North American MEMS Application Lab in Boston USA: For immediate release

2001-09-27

- Tronic's Microsystems has Leveraged 10,5 Million Euros to Increase its Manufacturing Capacities

2001-09-20

- Training Course: Start-up creation in the field of Microsystems: October 12 at Munich (DE)
- Training Course: CAD Tools for MEMS: October 8-9, 2001 at Munich (DE)

2001-09-19

- Advion BioSciences Selects IntelliSense to Manufacture ESI Chip?: Leading Mass Spectrometry Laboratory Transfers Microfluidic Chip Production to IntelliSense

2001-09-06

- Microsystems Technology Short Course: 28-30 November 2001 @ Imperial College, London, UK

2001-09-05

- Randall Wolf of Ohio State University joins Verimetra's Strategic Advisory Board

2001-08-29

- New Gold Deposition Service

2001-08-28

- David Hill Joins IntelliSense as President: Seasoned Semiconductor Executive takes Helm at Industry-Leading MEMS Company
- MEM Research released releases version 4.2 of EM3DS: A completely free version of the program is also available

2001-08-27

- IntelliSense and MM Solutions, Ltd. Partner to Develop European CAD for MEMS? Market: New Distribution partnership to strengthen support for European customers of IntelliSuite®

2001-08-08

- Expertech Announces New Reduced Footprint Line of Horizontal Furnaces Targeted for Emerging Device Market: Scotts Valley, CA., August 8, 2001

2001-08-06

- 2001 International Semiconductor Device Research Symposium (ISDRS): Sessions on MEMS Materials and Processes, Optical MEMS, Microfluidics and Power MEMS

2001-08-01

- Microtechnology at Hannover Fair 2002: USA to have greater presence

2001-07-31

- MEMS Technology Workshop: October 16-19, 2001, Boston, MA, USA

2001-07-27

- Verimetra appoints two leading surgeons to its Strategic Advisory Board

2001-07-26

- Ardesta Launches Discera: To Supply Miniaturized Communications Subsystems to Wireless Industry

2001-06-22

- IntelliSense Corporation co-sponsors Singapore MEMS course

2001-06-11

- microTEC rapid manufacturing without tooling
- ISSYS Inc. Receives $2.75 Million in Funding To Commercialize Advanced Micromachined Biofluidic Chips
- MEMSCAP Establishes Finnish Subsidiary for Active Optical Components
- MEMSCAP Offers Active Optical Component Products for POF
- PHS MEMS raises 31 million euros in a major funding round
- MEM Research has released EM3DS 4.0, a new software well suited for the electromagnetic analysis of MEM switches and capacitors

2001-06-01

- Verimetra, Inc. closes first round

2001-05-19

- IntelliSense Corporation co-sponsors Singapore MEMS course
- Fully automated wafer mounting and demounting

2001-04-29

- Walsin Lihwa Chooses MEMSCAP in Multimillion Dollar Partnership for MEMS Design and Production

2001-04-25

- Ardesta Announces Translume
- New Source for Daily Microsystems News goes Live: www.smalltimes.com
- Speakers list for MEMS Conference (Nick Ortyl ...)

2001-04-19

- New Conference and Trade Show Created for Emerging Small Tech Industry
- Kokusai Completes Customer Development and Demonstration of MEMS Thermal Processing Suite Using the Kokusai BDF-41 Horizontal Furnace Reactor

2001-04-14

- Tystar Secures Large Equipment Order For MEMS Device Production

2001-04-08

- Microtronic A/S (DK) and microFAB Bremen GmbH (D) join forces to set up production of silicon microphones

2001-03-16

- MEMSCAP Offers Integrated MEMS Design Platform for Cadence Users
- Berkeley MEMS Conference picking up steam

2001-03-03

- TRONICS Microsystems and DEBIOTECH close Collaboration Agreement

2001-02-22

- Mezzo Systems Announces Availability of LIGA Foundry Services
- IntelliSense Certified to ISO 9001 Quality Standard
- Conference on Health Engineering
- Nano Alumina Fiber -Product La\unch
- MEMSCAP'S MEMS COMM RF-MEMS LIBRARY AVAILABLE FOR POPULAR AGILENT TECHNOLOGIES ADS TOOL SUITE
- Umech Technologies Introduces First Commercial Full-Field 3D MEMS Motion Measurement Tool
- MEMSCAP EXPANDS OPTICAL MEMS PRODUCT PORTFOLIO
- XACTIX spins off bio-MEMS business
- DISCO IS AT THE CUTTING EDGE OF TINY TECHNOLOGY
- Microcosm Technologies and JENOPTIK Mikrotechnik team to support optical and microfluidics MEMS
- IEEE Starts Sensors Journal
- ARGONIDE ANNOUNCES NEW COPPER COATING PROCESS
- GT Equipment Technologies, Inc. (GTi) announces new website
- National Nanotechnology Initiative
- New paint capable of metallizing substrates with thick or thin films.
- ADC Teams with MEMSCAP for MEMS-Based Optical Communications Components
- NASA Hosts First IntelliSuite, CAD for MEMS, User Group Meeting
- Free Trial of CFD-Micromesh, a Fast 3D Model Builder from Layouts
- ELSEVIER launches SMART MATERIALS BULLETIN
- ANSYS, Inc. and MEMSCAP Announce Strategic OEM Agreement
- Next Generation Medical Products To Revolutionize Chronic Patient Monitoring: Implantable, Wireless, MEMS Sensing Systems

- Deutsche Effecten- und Wechsel-Beteiligungsgesellschaft AG is spinning off MILDENDO Gesellschaft fuer mikrofluidische Systeme mbH from JENOPTIK Mikrotechnik GmbH
- 300mm BREAKTHROUGH
- ARGONIDE WINS PRESTIOUS R&D 100 AWARD
- MicroTechnology: New display category at HANNOVER FAIR 2001
- JPSA ANNOUNCES THE IX-1000 LASER MICRODICE SYSTEM:
- MEMSCAP SECURES $11M IN FUNDING TO BROADEN MEMS COMMERCIAL DESIGN TECHNOLOGY
- MEMSCAP RF-MEMS DEVICE LIBRARY ENABLES SINGLE-CHIP WIRELESS DESIGN
- The next generation of software from Atomasoft
- MICROCOSM TECHNOLOGIES, INC. ACQUIRES COYOTE SYSTEMS, INC.
- MICROCOSM TECHNOLOGIES, INC. EXPANDS SYSTEM LEVEL SUPPORT FOR TELECOMMUNICATIONS MEMS DEVELOPMENT.

5

Nanoengineering & Nanotechnology: Industrial & Other Applications

Molecular Mechanics

Molecular mechanics consists of simulating the behavior of an object (e.g. a nanotube) by calculating the interaction of every atom with its neighbors. For example, this powerful technique enables us to calculate how tubes of different diameter are distorted when placed on a substrate. It costs the tubes energy to bend or deform, but in turn they can gain energy from the van der Waals interaction with other tubes or with the substrate they are deposited upon.

The figure above shows the radial deformation of adsorbed single-wall carbon nanotubes calculated using molecular mechanics. The extent of radial deformation increases as the tube diameter is increased. The system can lower its energy by increasing the area of contact between the tube and the substrate. We have simulated the situation where two adsorbed nanotubes cross each other at a right angle. The bottom nanotube is compressed by the top tube which is strongly attracted to the substrate. From these calculations we can compute the force exerted on the lower tube, and understand how nanotubes deform.

Molecular Logistics - Teleportation and Molecular Nanotechnology

Most all of us remember the Transporter on Star Trek and the "beam me up Scotty" phrase that typified use of this particle beam transportation technology. Most all of you will also probably scoff at this as a topic for a serious Logistics article, but following our recent review of this topic and the related field of Molecular Nano-Technology, we thought an introduction and/or update about them for our members was in order. And with the millenium only five years old, we also see this as an opportunity to introduce technologies we believe will be to the 21st Century what the truck, tractor trailer, intermodalism and forklift were to this century.

The basic concept of Teleportation is that a physical object would get broken down to its' component molecular parts at one location and then transmitted to another location where they would be reassembled back to the original form completing the process of transport without any of the current costs, equipment or time delays.

A listing of a number of sites of interest relative to Teleportation can be found below;

- Quantum Teleportation across the Danube Demonstrated
- Teleportation Takes Quantum Leap
- IBM Research Quantum Teleportation
- American Institute of Physics—Quantum Teleportation

Molecular Nano-Technology is a slightly different although similar approach whereby the required molecules for the creation of a specific item are precisely assembled to "build" the item at or near the location of consumption.

This takes the concept of Teleportation back even one step further to actually eliminate current production processes. In fact in most cases production today is based on our taking matter and cutting/forming it into the product required. Molecular Nano-Technology uses the reverse of this concept where the intent is to microscopically place through exact placement precisely only the molecules required to create the given item.

Again some sample pages from across the net are presented for your information in regards to molecular Nano-Technology;

- Institute for Molecular Manufacturing
- Foresight Institute Update on Nanotechnology
- Senate Hearings- Molecular Nanotechnology

Finally, before you jump all over our review of these two futuristic technologies, please suspend your disbelief long enough to review some of the links above focussed on identifying the current status in both cases.

Although it will be many years before commercial applications of these technologies emerge, developments are now coming fast and furious. It is important for logisticians to stay abreast of these breakthroughs and provide industry input into the development of nanotechnology and teleportation.

Novel Self-Assembly Processes For Nanotech Applications

Researchers at the University of Massachusetts Amherst have developed a series of novel techniques in nanotechnology that hold promise for applications ranging from highly targeted pharmaceutical therapies, to development of nutrition-enhanced foods known as "nutraceuticals," to nanoscopic sensors that might one day advance medical imaging and diagnostics. The research, published in the Jan. 10 issue of Science, was funded by the U.S. Department of Energy and the National Science Foundation. The team included faculty members Thomas Russell and Todd Emrick of the department of polymer science and engineering, and Anthony Dinsmore of the department of physics, and graduate students Yao Lin and Habib Skaff, both of polymer science and engineering. "Our findings open new avenues to revolutionize technology by the controlled fabrication of nanoscopic materials having unique optical, magnetic and electronic properties," said Russell.

The study details three major findings:

- A novel method to create robust capsules from nanometer-sized particles
- A new technique to make nanoscopic particles water-soluble
- Functionalizing regions of the capsules with tailored properties, such as luminescence.

Emrick's research explores the behavior of nanoparticles to which ligands organic molecules and polymers have been attached. Russell is an expert in the surface and interfacial properties of polymers, and polymer-based nanostructures. Dinsmore specializes in colloidal assemblies and interface physics. "This is a productive collaboration in that we really have all the bases covered in terms of synthesis, understanding of interfacial activity and mediation, and the physics issues including surface tension and particle interactions," Emrick said. The study details a new method for assembling nanoparticles into robust, three-dimensional structures by encapsulating and stabilizing water droplets. Nanoparticles suspended in oil will self-assemble around a droplet of water, fully coating it with a shell. Although scientists have long known that particles tend to assemble at fluid interfaces, "the idea of using liquid interfaces as scaffolds is exciting and tremendously useful since researchers can tailor or modify the nanoparticles from both sides of the interface," explained Dinsmore. "We have much more surface area to work with for adding or removing specific particles."

"Nanoparticles have exciting properties due to their small size, and they can be prepared in various shapes and sizes. What's really key is that you attach ligands that extend from the nanoparticles like hairs, in order to preserve the nanoscopic integrity of the particles and prevent them from clustering," Emrick said. "Changing the nature of these organic ligands can really modify the behavior of the particles. You can endow the nanoparticles, and thus the capsules that they form upon interfacial assembly, with a wide range of properties based on which ligands are attached." The effect of the ligands on the interactions of nanoparticles with the surrounding environment is crucial in medical applications. "These organic molecules will dictate the solubility, miscibility, and charge transport properties of the particles," Emrick said.

UMass researchers also developed a method to take these nanoparticles, which are oil-soluble, and make them water-soluble, simply by shining light on them. "Developing nanoparticles that are water-soluble has significant implications for medicine in the biosensors area," Emrick said. "Using luminescent material, as we did, could lead to advances in very sophisticated medical-imaging techniques as the fluorescent nature of these particles allows them to be viewed and tracked over time."

Finally, the UMass team discovered that when nanoparticles of different sizes compete for assembly at the interface, the bigger ones win, and segregate or cluster into patches on the droplet surface. "This opens a range of possibilities for developing nanoscopic capsules that have certain properties in specific areas," said Dinsmore. "You could build in an area with permeability, magnetism, or conductivity, so that one area would be functionally distinct."

In just a few decades physicians could be sending tiny machines into our bodies to diagnose and cure disease. These nanodevices will be able to repair tissues, clean blood vessels and airways, transform our physiological capabilities, and even potentially counteract the aging process. It is the year 2031, and the age of advanced nanomedicine has arrived. A young man arrives at his physician's office with a mild fever, nasal congestion, discomfort and a cough. The physician pulls from her pocket a lightweight, handheld device resembling a pocket calculator. She unsnaps from it a cordless, self-sterilizing, pencil-sized probe and inserts it into the patient's mouth as if it were a tongue depressor. On the tip of the probe are billions of molecular assay receptors, mounted on hundreds of self-guiding retractile stalks. Each receptor is sensitive to the chemical signature of a specific kind of bacterium or virus. "Ahhh," says the patient, and a few seconds later a three-dimensional, color-coded map of his throat appears on the display panel of the device. Beneath the map scroll columns of data, revealing the unique molecular signature of a known and unwelcome bacterial pathogen. With the diagnosis complete, the infectious microorganism can be exterminated. No need for antihistamines, cough drops and a week-long course of antibiotics. The physician keeps several generic classes of nanorobots in her office for just such a circumstance. Using a desktop appliance in her office, she programs billions of nanorobots to find, recognize and destroy the particular microbial strain. The nanomachines are suspended in a carrier fluid that the patient inhales into his lungs, after which the mobile devices march down the patient's throat, propelled on tiny legs. Following a search pattern, the nanorobots ingest and destroy the harmful bacteria they encounter using mechanical and chemical phagocytosis. The patient feels nothing: nanorobots are the size of bacteria, which constantly crawl on and inside the body without ever being noticed. After several minutes, the physician activates an acoustic homing beacon to guide the nanorobots back into the patient's mouth, where she retrieves them through a collection port on the tip of the homing device. A further survey with the original diagnostic probe reveals no evidence of the pathogen. This is a wonderful vision of medicine in the future, but how do we get there from here? Nanotechnology ("nano" from the Greek word for dwarf) involves engineering and manufacturing at the molecular, or nanometer, level. One nanometer is one billionth of a meter, about the width of six bonded carbon atoms. How does one build at this scale? Scientists at New York University have adopted the self-assembly approach, and have succeeded in producing complementary strands of DNA that can zip themselves into complex structures. They have built cubes, octahedra and other solids made of just a few thousand nucleotides each, by the billions per batch. Meanwhile at Cornell University, researchers have genetically engineered a natural biomotor normally found in the enzyme ATPase to allow it to incorporate nonbiological parts such as a silicon nitride bar 100 nanometers in length, making the first artificial hybrid nanomotor. In a microscopic video presentation, dozens of bars attached to artificial molecular motors in a large precise array could be seen spinning at 200 revolutions per minute, like a field of tiny propellers.

Another way to build at the molecular level is by positional assembly—picking and placing molecular parts exactly where you want them. A device capable of positional assembly would work much like the robot arms that manufacture cars on automobile assembly lines, or like the ribosome that assembles amino acids one by one into proteins in our cells. The US engineering firm Zyvex Corp. is aiming to be the first to create an artificial molecular assembler using positional assembly to manufacture atomically precise structures.

Elsewhere, researchers at Harvard University have constructed the first general-purpose "nanotweezers," using a pair of electrically controlled carbon nanotubes These have successfully grasped 300-nanometer clusters of polystyrene spheres and extracted a single semiconductor wire 20 nanometers wide from a mass of entangled wires. The scientists hope eventually to produce nanotweezers small enough to grab individual large molecules. Such nanoresearch will eventually lead to applications affecting almost every aspect of modern life, including the home, transport, computers, energy and the environment. Some of the most important and life-changing applications, however, are likely to be seen in medicine. We may see the first nanomedical materials and devices in use within the next few years. Scientists at the University of Michigan's Center for Biologic Nanotechnology are currently pursuing the use of dendrimers as a safer and more effective genetic therapy agent. Dendrimers are branching synthetic molecules that can be grown nanometer by nanometer to reach the desired size. They take on a spherical shape, and have sufficiently large openings and cavities to carry small molecules—dendrimers could therefore be used to sneak DNA into cells without triggering an immune response. At the same center, researchers have reported using dendrimer "nanodecoys" to trap and deactivate influenza virus particles. Relatively simple nanodevices could soon offer cures for major conditions such as diabetes. Researchers at Ohio State University and the University of Illinois have managed to construct silicon-based microcapsules highly perforated with "nanopores," each as small as 20 nanometers in diameter. These pores are large enough to allow small molecules such as oxygen, glucose and insulin to pass through, but are small enough to impede the passage of much larger immune system molecules such as immunoglobulins. Microcapsules containing islet cells could be implanted beneath the skin of some diabetes patients, temporarily restoring the body's delicate glucose control feedback loop without the need for powerful immuno-suppressants that can leave the patient at serious risk for infection. The same method of encapsulation could be used to treat other enzyme- or hormone-deficiency diseases. Encapsulated neurons could also be implanted in the brain and then electrically stimulated to release neurotransmitters, possibly as part of a future treatment for Alzheimer's or Parkinson's disease.

Artificial "biobots" could be in our bodies within five to ten years. Advances in genetic engineering are likely to allow us to construct an artificial microbe—a basic cellular chassis—to perform certain functions. These biobots could be designed to produce vitamins, hormones, enzymes or cytokines in which the host body was deficient, or they could be programed to selectively absorb and break down poisons and toxins. A new company called EngeneOS, Inc., founded in late 2000, has already announced plans to develop artificial Engineered Genomic Operating Systems using the techniques of molecular biology. These systems will comprise a library of component device modules and proprietary modular components. This will allow the engineering and construction of programmable biobots with novel form and function. The greatest power of nanomedicine will emerge in the longer term, perhaps 10-20 years from now, when we learn how to design and construct complete artificial nanorobots using strong diamond-like materials, nanometer-scale parts, and onboard subsystems including sensors, motors, manipulators, power plants, and molecular computers. One example is an artificial mechanical cell called a respirocyte, which could be used to keep a patient's tissues safely oxygenated for up to about four hours (at maximum dosage) if their heart has stopped beating. These cells (side panel, 4) could also enable a healthy person to sprint at top speed for at least 15 minutes without breathing, or to sit underwater at the bottom of a swimming pool for hours. Still entirely theoretical, the respirocyte is a micron-wide spherical nanorobot made of 18 billion atoms precisely arranged in a diamondoid structure to form a tiny 1,000-atmosphere pressure tank. Several billion molecules of oxygen and carbon dioxide can be absorbed into, or released from, this tank using computer-controlled molecular pumps powered by serum glucose and oxygen. External gas concentration sensors would allow respirocytes to mimic the action of the natural hemoglobin-filled red blood cells, with oxygen released and carbon dioxide absorbed in the tissues, and vice versa in the lungs. Each respirocyte would be able to hold 200 times more gas per unit volume than a natural red cell, so a few cubic centimeters injected into the human bloodstream would exactly replace the gas carrying capacity of the patient's entire 5.4 liters of blood.

Other proposed medical nanorobots offer equally astonishing performance improvements over nature. For instance, micronsize artificial mechanical platelets could allow complete hemostasis (control of bleeding by clot formation) in just one second, even for moderately large wounds, a response time 100-1,000 times faster than the natural system. This would be achieved by rapidly unfurling a compactly stowed onboard biodegradable mesh under the control of an onboard computer. These "clottocytes" would be about 10,000 times more effective as clotting agents than an equal volume of natural platelets. Nanorobotic phagocytes called microbivores could patrol the bloodstream, seeking out and digesting unwanted pathogens including bacteria, viruses or fungi. Each nanorobot could completely destroy one pathogen in just 30 seconds—about 100 times faster than natural leukocytes or macrophages—releasing a harmless effluent of amino acids, mononucleotides, fatty acids and sugars. No matter that a bacterium has acquired multiple drug resistance to antibiotics or to any other traditional treatment. The microbivore will eat it anyway, achieving complete clearance of even the most severe septicemic infections in minutes to hours, as compared to weeks or even months for antibiotic-assisted natural phagocytic defenses, without increasing the risk of sepsis or septic shock. Related nanorobots could be programed to recognize and digest cancer cells, or to clear circulatory obstructions within minutes in order to rescue stroke patients from ischemic damage. More sophisticated medical nanorobots will be able to intervene at the cellular level, performing surgery within cells. Physician-controlled nanorobots could extract existing chromosomes from a diseased cell and insert newly manufactured ones in their place, a process called chromosome replacement therapy. This would allow a permanent cure of any pre-existing genetic disease, and permit cancerous cells to be reprogramed to a healthy state.

In recent years, many gerontologists have begun to think of aging as a genetic disease that might be cured. Evidence from the Human Genome Project suggests that just a few hundred genes at most may be directly involved in aging. If we can completely understand these few genes and how they work, we may be able to alter them to eliminate this unwanted syndrome. Using cytosurgical nanorobots, corrected genes could be installed in every one of the 10 trillion tissue cells in our bodies. We would then no longer naturally age, and our bodies would again repair themselves as well as they did when we were children. Although nanotechnology is in its infancy, researchers are steadily making major breakthroughs. If we can learn to harness and precisely control the ability to manipulate molecules, then many aspects of our lives will change forever. In particular, the ability to carry out medical procedures at the molecular level will revolutionize medical practice. The next few decades will be very interesting indeed.

A new era on medicine are expected to happen in the coming years. Due to the advances in the field of nanotechnology, nanodevice manufacturing has been growing gradually. From such achievements in nanotechnology, and recent results in biotechnology and genetics, the first operating biological nanorobots are expected to appear in the coming five years, and more complex diamondoid based nanorobots will become available in about ten years. In terms of time it means a very near better future with significant improvements in medicine. In this work we present a practical approach taken on developing nanorobots for medicine in the sense of using computational nanomechatronics techniques as ancillary tools for investigating manufacturing design, nanosystems integration, sensing and actuation for medicine applications. Thus, the work describes pathways that could enable design testability, but also help scientists and profit corporations in providing the helpful information needed to test and design integrated devises and solutions towards manufacturing biomedical nanorobots. The use of robots in surgery has provided additional tools for surgeons enabling minimally invasive intervention or even long distance teleoperated surgeries. Indeed we may trust on human creativeness and technical capabilities that can ever be improved in terms of technical achievements. In recent years the medicine has enabled significant wellness for the life quality and longevity of the world population. And for the coming years, we may be prepared to experiment even more benefits, as results from advances that are being pursued step by step in new fields of science, such as nanobiotechnlogy. With the expected miniaturization of devices provided by several works on nanoelectromechanical systems (NEMS), nanomanufacturing has actually become a reality. Hence, with the NEMS recent advances on building nanodevices, and the development of interdisciplinary works, altogether may be translated in few years through the development of integrated nanomachines, also known as nanorobots. With the use of techniques that are advancing rapidly, such as nano-transducers, and biomolecular computing, nanorobots are expected to be able to operate in a well defined set of behaviors performing pre-programed tasks. Thus, in the coming few years, nanorobots being teleoperated to perform surgery, or even nanorobots continually supervising the human body in order to assist organs that may require some kind of repair, is one of the most expected revolutionary tools for biomedical engineering problems.

The development of nanorobots is an emerging field with many aspects under investigation. Simulation is an essential tool for exploring alternatives in the organization, configuration, motion planning, and control of nanomachines exploring the human body. Nanorobot applications could be focused mainly on two major areas, as follows: nanorobots for surgical interventions, as well as their utilization for patients that need constant monitoring. The nanorobots require specific controls, sensors and actuators, basically in accordance with each kind of biomedical problem. Advanced simulations can include various levels of detail, giving a trade-off between physical accuracy and the ability to control large numbers of nanorobots over relevant time scales with reasonable computational effort. Another advantage is that simulation can be done in advance of direct experimentation. It is most efficient to develop the control technology in tandem with the fabrication technologies, so that when we are able to build these devices, we will already have a good background in how to control them.

Computational mechatronic approaches were proposed as a suitable way to enable the fast development of nanorobots operating in a fluid environment relevant for medical applications. Unlike the case of larger robots, the dominant forces in this environment arise from viscosity of low Reynolds number fluid flow and Brownian motion and such parameters are been implemented throughout a set of different investigations. Practical and innovative paradigms have been developed based on the Nanorobot Control Design (NCD) simulator that allows fast design testability comparing various control algorithms for nanorobots and their application for different tasks. Also such information generated by the NCD can be useful as parameters for building nanodevices, such as transducers and actuators.

In future decades the principal focus in medicine will shift from medical science to medical engineering, where the design of medically-active microscopic machines will be the consequent result of techniques provided from human molecular structural knowledge gained in the 20th and early 21st centuries. For the feasibility of such achievements in nanomedicine, two primary capabilities for fabrication must be fulfilled: fabrication and assembly of nanoscale parts. Through the use of different approaches such as biotechnology, supramolecular chemistry, and scanning probes, both capabilities had been demonstrated to a limited degree as early as 1998. Despite quantum effects which impart a relative uncertainty to electron positions, the quantum probability function of electrons in atoms tends to drop off exponentially with distance outside the atom. Even in most liquids at their boiling points, each molecule is free to move only ~0.07 nm from its average position. Developments in the field of biomolecular computing have demonstrated positively the feasibility of processing logic tasks by biocomputers, a promising first step toward building future nanoprocessors with increasing complexity. There has been progress in building biosensors and nanokinetic devices, which also may be required to enable nanorobotic operations and locomotion. Classical objections related to the feasibility of nanotechnology, such as quantum mechanics, thermal motions and friction, have been considered and resolved and discussions of techniques for manufacturing nanodevices are appearing in the literature with increasing frequency.

One important challenge that has become evident as a vital problem in nanotechnology industrial applications is the automation of atomic-scale manipulation. The starting point of nanotechnology to achieve the main goal of building systems at the nanoscale is the development of control automation for molecular machine systems, which could enable the massively parallel manufacture of nanodevice building blocks. Governments all around the world are directing significant resources toward the fast development of nanotechnology. At least 30 countries have initiated activities in this field, and beyond that, with government investments to nanotechnology of US$ 800 million in Japan and US $ 774 million in USA. The U.S. National Science Foundation has launched a program in "Scientific Visualization" in part to harness supercomputers in picturing the nanoworld. A US $ 1 trillion market consisting of devices and systems with some kind of embedded nanotechnology is projected by 2015. More specifically, the firm 'display search' predicts rapid market growth from US $ 84 million today to $ 1.6 billion in 2007. A first series of commercial nanoproducts has been announced as foreseeable by 2007. To reach the goal of building organic electronics, firms are forming collaborations and alliances that bring together new nanoproducts through the joint efforts of companies such as IBM, Motorola, Philips Electronics, Xerox/PARC, Hewlett-Packard, Dow Chemical, Bell Laboratories, and Intel Corp., among others. For such goals, new methodologies and theories to explore the nanoworld are the key technology.

A useful starting point for achieving the main goal of building nanoscale devices is the development of generalized automation control for molecular machine systems which could enable a manufacturing schedule for positional nanoassembly manipulation. In our work we consider more specialized scheduling problems with a focus on nanomedicine: describing in a detailed fashion the nanorobot control designs and the surrounding virtual workspaces modelling that are required for the main kinematics aspects in the physically-based simulations. Here the biomolecular assembly manipulation could be automatically performed by smart agents, which are given a set of possible tasks for biomedical engineering problems, embedded in a complex 3D environment. Virtual Reality could be considered as a suitable technique for nanorobot design and for the use of macro and microrobotics concepts given certain theoretical and practical aspects that focus on its domain of application. The collective nanorobotics approach proposed is one possible method to perform a massively-parallel positional nanoassembly manipulation . We constructed and demonstrated the applicability of multi-robot teams in timely sequenced set of works with practical applications that could enable the establishment of generalized control guidelines for nanorobotics. The use of multirobot teams working cooperatively to achieve a single global task applied to nanotechnology is a field of research that is very new and challenging. Research on collective robotics suggests that we should consider emulating the methods of the social insects to build decentralized and distributed systems that are capable of accomplishing tasks through the interaction of agents with the same structures and pre-programed actions and goals. Thus, a careful decomposition of the main problem task into subtasks with action based on local sensor-based perception could generate multirobot coherent behaviors. Several techniques was applied for such aims, as Neural Networks algorithms, Evolutionary computation, chemical based sensors and actuators, and even temperature time-gradients, just to quote a few. Among other interesting nanorobot applications, we could foresee their use to process specific chemical reactions in the human body as ancillary devices for injured organs. Nanorobots equipped with nanosensors could be used to detect glucose demand in diabetes patients. Moreover, nanorobots could also be applied in chemotherapy to combat cancer through superior chemical dosage administration, and a similar approach could be taken to enable nanorobots to deliver anti-HIV drugs. Successful nanorobotic systems must be able to respond efficiently in real time to changing aspects of microenvironments not previously examined from a control perspective. Unlike some prior simulators for simple robots, in the present work we have developed nanorobots that are not restricted to a fixed grid nor behave as simple cellular automata with very simple environments. Also by contrast, most CAD approaches provide only animation or visualization tools, while the NCD is a physically-based simulator. The set of experiments we have carried out include the main physical properties existent in the environment where the nanorobots are being currently projected to be operating in the coming years. Hence, the majority of results from our investigation should be useful as well on analysing integrated manufacturing capabilities. In fact, including aspects of the physical environment in conjunction with graphical visualization provide a feasible approach for automation and control design. Furthermore, the architecture that was developed intended to enable the incorporation and evaluation of several control methods and distinct nanorobot shapes analyses. The automation, control, and manufacturing of nanorobots is a challenging and very new field. Realizing revolutionary applications of nanorobots to health or environmental problems raises new control challenges. The design and the development of complex nanomechatronic systems with high performance should be addressed via simulation to help pave the way for future applications of nanorobots in biomedical engineering problems.

Soft Nanotechnology and Self-Assembly: Industrial Applications

The focus of this course is the design and application of nanostructured fluids and soft materials, with a particular emphasis on self-assembly processes. Many soft or fluid consumer products, such as foods, paints, detergents, personal care products, and cosmetics contain nanometer to micron scale structures. These structures are formed by the spontaneous or directed self-assembly of surfactants and polymers. In many cases complex mixtures are required to create the desired structures and performance. These materials are difficult to formulate, and challenging to characterize; but scientists in these industries, and in the academic groups focusing on these areas, have developed considerable expertise in this area. New insights and techniques have the potential to revolutionize product design. This knowledge is being applied in new areas such as drug encapsulation, solubilization and delivery, in oil recovery, environmental remediation, and in new "active" cosmetics, nutraceuticals, antimicrobials and smart materials.

Members of the more broadly defined "nanotechnology community" are also increasingly interested in self-assembly and nanostructured fluids because of their potential to provide robust and inexpensive strategies for creating nanoscale materials. For example, IBM recently highlighted the potential for block-copolymer self-assembly to create nanometer size structures for chip fabrication, and soft nanostructured materials can provide templates for synthesis of nanoparticles. Graduates of this course will leave with a broad appreciation for current and future applications of soft nanostructured fluids, soft materials, and of self-assembly. They should have gained a sufficient grasp of the concepts and language to read and understand most research papers in these areas. They will be alerted to the latest developments in this fast moving field, and to approaches for designing materials for particular applications. They will have been introduced to characterization and to computer-modeling techniques appropriate for these materials. They will have gained insight into the surprising commonality to design of some very different products, and also the way soft materials are starting to be used to create "hard" nanostructures.

Nano-Vasculoid

As outlined below, at peak average male human metabolic rates, ~5400 cm^3 of circulating blood transports up to ~125 cm^3 of essential molecules and up to ~90 cm^3 of essential cellular elements (excluding erythrocytes). Consequently ~96 per cent of blood volume (mostly water) may be rendered superfluous if its salvation and suspension functions can be replaced with a more efficient and highly reliable nanomechanical transport mechanism. The resulting baseline vasculoid device conveys physiologically useful molecules and cells throughout its interior, and thence to the biological tissues, by containerizing all materials and then using a ciliary transport system to distribute the containers to appropriate destinations. This architecture thus utilizes the fractal branching vascular network that is characteristic of efficient fluid transport systems, although many architectures are possible in this design space.

In natural human physiology, the circulating blood moves ~2 x 10^{26} molecules/sec. However, most of these molecules are solvent water not strictly necessary in human metabolism. In the vasculoid, all non-cell materials transport takes place via ciliary transport of tanker vessels at a very conservative mean transport velocity of 1 cm/sec. (For comparison, blood velocity in the natural vasculature ranges from 0.02-0.15 cm/sec in the capillaries up to 117.5 cm/sec pulses in the femoral artery). Tanker vessels are $(1\text{-micron})^3$ mostly-sapphire (possibly chamfered) cubes with 0.75 $micron^3$ useful interior storage volume and ~8 × 10^{-16} kg dry mass. Gases are stored in tankers at 1000 atm pressure; this is highly conservative, as a sapphire pressure vessel should easily withstand ~10^5 atm without bursting.. Liquids and solids are packed at normal macroscopic densities. All tanker mechanical subsystems (other than passive hull) have at least tenfold redundancy, hence are extremely reliable. The following is an approximate inventory and assessment of the minimum human physiological molecular transport requirement.

The human body consumes 1-20 × 10^{20} molecules/sec of O_2, depending on activity level. Similar numbers of CO_2 molecules are generated as waste products. The human red blood cell mass (~2.4 liters of RBCs) has a total storage capacity of 3.2 × 10^{22} O_2 molecules, but since hemoglobin operates between 70-95 per cent saturation in normal physiological conditions, the active capacity of human blood is only 8.1 × 10^{21} O_2 molecules. Blood CO_2 capacity is roughly the same, and O_2/CO_2 are loaded/unloaded reciprocally at lungs and tissues. Packing density of CO_2 molecules (1.11 × 10^{28} molecules/m^3) is slightly lower than for O_2 molecules (1.26 × 10^{28} molecules/m^3) at 1000 atm, so CO_2 packing density controls respiratory tanker population. Assuming a mean circulatory circuit of ~1.4 meter and a 1 cm/sec transport speed, continuous transport at the maximum 20 x 10^{20} CO_2 molecules/sec rate requires a circulating fleet of 33.6 trillion respiratory gas tankers. At normal basal metabolic rates, only 1.68 trillion active tankers are required; the remainder are held in reserve to support periods of more intense physical activity. Each tanker can hold 9.48 x 10^9 oxygen molecules or 8.36 x 10^9 molecules of carbon dioxide at 1000 atm pressure. Highly efficient reciprocal regenerative energy recovery is assumed during gas loading and unloading operations and unrecoverable gas compression energy losses should be minimal. Isothermal compression of ~10^{10} O_2 or CO_2 molecules requires—P_i V_i $\ln(V_f/V_i)$ ~ (10^5 N/m^2) (0.75 x 10^{-18} m^3/tanker) ln(1/1000) = 0.5 pJ/tanker. Assuming a need for 2 x 10^{20} molecules/sec for O_2 and CO_2 at basal demand implies ~2 x 10^{10} tanker fills/sec or ~0.01 watt for both sets of gases, or up to ~0.2 watts at peak demand, for compression of gases. Conservatively taking the CO_2 concentration in the vasculoid-endothelium interface fluid as equal to the normal CO_2 venous plasma concentration of C = 0.45-1.14 x 10^{24} molecules/m^3, then the diffusion limit for loading CO_2 from the tissues to one side of a square plate area of L^2 = 2 micron^2, with the diffusion coefficient D = 1.9 x 10^{-9} m^2/sec for CO_2 at 310 K, is J = ($4/p^{1/2}$) LDC = 2.7-6.9 x 10^9 molecules/sec; a 10-second fill time allows system operation at about one order of magnitude below the diffusion limit. Similarly, normal alveolar partial pressure of O_2 is >100 mmHg, or C ~ 3.1 x 10^{24} molecules/m^3; taking D = 2.0 x 10^{-9} m^2/sec for O_2 at 310 K, the diffusion limit for loading oxygen at the lungs is J ~ 2 x 10^{10} molecules/sec, which again allows system operation at about one order of magnitude below the diffusion limit assuming a 10-sec tanker fill time. Tankers are used to transport O_2 and CO_2, rather than the vasculoid interior in part because the theoretical rupture pressure of the natural cylindrical aorta is p_{max} ~ 1 atm (taking aortic wall thickness t_{wall} = 1.5 mm, aortic radius R = 12.5 mm, and working stress s_w ~ 10^6 N/m^2 and the rupture pressure of the vasculoid plate coating is p_{max} ~ 8 atm (taking vasculoid plate tube thickness t_{wall} = 1 micron, tube radius R = 12.5 mm, and working stress s_w ~ 10^{10} N/m^2. At peak exertion levels, respiratory gases stored in the tanker fleet at 1000 atm would fill the 4.2 liters of free vasculoid internal volume to 3.2 atm of pressure, too high for reliable safe operation under all foreseeable circumstances. By comparison, the rupture strength of individual tankers exceeds ~40,000 atm (the rupture strength of individual plates).

The human body excretes 6-12 x 10^{20} molecules/sec of urinary water under normal circumstances. The kidney continuously filters ~18 gallons/hour of blood (~7 x 10^{23} H_2O molecules/sec) so much higher rates of urine generation are possible in theory, but it seems unnecessary to accommodate the vasculoid design to such extreme cases because the removal of waste molecules can be accomplished by direct molecular sorting rotor extraction rather than by bulk filtration and solvent repumping—that is, the principle blood-cleansing function of the kidneys is to pump nonaqueous molecules, not bulk water solvent molecules. With the vasculoid in place, only specific waste molecules and minimal quantities of carrier water need be discharged into the glomeruli of the ~10^6 nephrons of the kidney, with the result that gross water flows through the kidney and energy consumption in this biological organ can be significantly reduced—but not reduced so far as to induce urinary stone disease (urinary lithiasis or urolithiasis) or renal cytotoxicity from excessive pH, salinity (e.g., hyperosmotic stress), or toxin concentrations. The precise volume and concentration of urine generated by the biological organ is regulated by hormones that control kidney function, and these hormones may in turn be regulated by the vasculoid appliance. Additionally, 2-186 x 10^{20} molecules/sec of water are perspired through the skin, depending upon air temperature, humidity, activity level and mental state. Assuming the same transport parameters as before, the maximum requirement of 2.0 x 10^{22} H_2O molecules/sec (~0.6 cm^3/sec) requires 110.5 trillion water tankers to be available, of which only 4.5 trillion will normally be active with the rest held in reserve for periods of peak exertion. Each tanker can hold 2.51 x 10^{10} water molecules in the liquid state.

Typically 1.6 x 10^{22} molecules of glucose are present in the human bloodstream. A human on a standard 2000 kcal/day diet metabolizes 3.08 x 10^{19} glucose molecules/sec throughout the body. However, during strenuous exercise this requirement can briefly rise to as much as 33,000 kcal/day, a 4.93 x 10^{20} molecule/sec rate. Assuming the usual ciliary transport parameters, we require 17.6 trillion glucose tankers, providing storage of 6.91 x 10^{22} glucose molecules in the system, four times the number present in normal blood. (Slightly smaller capacity might be realized if the glucose must be stored as a 70 per cent aqueous solution concentrate to facilitate molecular mobility for sorting rotors. Only 1.1 trillion glucose tankers are needed to service basal metabolic needs; the remainder are held in reserve for periods of extreme physical activity. The average human cell is a ~30 picowatt (basal) device, so each glucose tanker, with its full complement of 3.93 x 10^9 glucose molecules, yields ~400 basal cell-seconds of energy (assuming 65 per cent efficiency for the natural cellular metabolic pathways using glycolysis and tricarboxylic acid cycles). Note that at higher power levels, the delivery rate increases correspondingly, so that even with random delivery at peak demand all cells should receive a sufficient fuel supply. Additionally, the average tissue cell normally contains an internal inventory of ~3 x 10^{10} glucose molecules, an in cyto buffer supply equivalent to ~8 full tanker loads.

Normal bloodstream concentrations of urea, uric acid, creatinine, creatine, ammonium salts, cerebroside, amino acids, etc. imply 1.73-2.49 gm present in the circulation and a maximum of 2.3 trillion tankers for full encapsulation and transport.

All plasma proteins including most notably albumen, globulins (including antibodies), complement, fibrinogen and prothrombin constitute ~7 per cent of adult blood plasma, or 210 gm, which would require 215 trillion tankers to encapsulate. Fortunately, on a transactional basis this figure is a gross overestimate. The RDA (Recommended Daily Allowance) active requirement for protein is ~70 gm/day for a 70 kg male human, an 8.1 x 10^{-7} kg/sec transport demand requiring only 0.12 trillion protein tankers for continuous ciliary transport, reducing mandatory instantaneous system storage capacity to ~0.12 gm of plasma proteins. However, almost the entire RDA protein intake is hydrolyzed to amino acids and dipeptides, so the actual protein transport needs should be far less. With no need to maintain osmotic pressure, there would be almost no reason to transport albumin within the system, although the dosage of protein-bound medication and hormones might need to be corrected. Finally, there is sufficient tanker capacity to increase protein transport capacity at least 100 fold above the RDA requirement if necessary, so the design is not particularly sensitive to this transport requirement. Note also that the five antibody classes can be efficiently transported using just five types of sorting rotors having binding pockets complementary only to the constant Fc regions without regard to the numerous variable domains these antibodies may encode on their antigen-binding Fab regions, thus vastly reducing the required vasculoid rotor specificities needed to accomplish this task. Complement (immune system) components may be transported with similar efficiency. The limited free enzyme activities that occur in natural blood can still take place in the fluids at the vasculoid-endothelial interface, with due care taken to avoid inducing localized hyperenzymemia (which is often benign, as in hypertransaminasemia), although higher concentrations may be expected because of the lower total volume of the vasculoid-endothelial interface compared to the natural blood volume.

The mass of lipids including fatty acids, cholesterol, triacylglycerides (transported in the plasma lipoproteins as triacylglycerols), and phospholipids present in the human bloodstream totals roughly 35-42 gm, which would require 55-66 trillion tankers to encapsulate. However, the RDA for fat is 23 per cent of caloric intake. (Many people consume several times this amount, but others survive on fat-free diets, so the figure represents a reasonable compromise estimate of the typical human active requirement.) This implies a demand of 6.9×10^{-7} kg/sec for a 2000 kcal/day diet, requiring only 0.15 trillion lipid tankers for continuous ciliary transport and reducing the required vascular storage capacity to ~0.1 gm. As with plasma proteins, there is sufficient tanker capacity to increase lipid transport capacity to at least 100 times the RDA requirement, so once again the design is not particularly sensitive to this transport requirement.

Human blood plasma contains ~40 gm of salts (e.g., electrolytes) including sodium, potassium, calcium, magnesium, chloride, phosphate, and sulfate; ~1.2 gm lactic acid and other non-nitrogenous waste products; ~0.23 gm of RNA and DNA; ~0.21 gm of enzymes (protein); ~0.15 gm testosterone in the male, ~0.03 gm progesterone in the female, ~0.02 gm of corticosteroids, and a total of ~0.002 gm of more than fifty other major biologically active hormones; ~0.04 gm of vitamins including A, B2, niacin, pantothenic acid, biotin, pteroylglutamic acid, choline, inositol, C, D, and E; ~0.03 gm of essential trace elements including iron, zinc, copper, manganese, iodine, and cobalt; and smaller masses of various cell-signaling (e.g., cytokines, autocrines, neuropeptides) and numerous other specialty proteins. Other nonprotein regulatory molecules affecting numerous signaling pathways must also be locally regulated with precision by the vasculoid appliance. For example, in the present context the simple molecule NO is of particular importance as a neurotransmitter and vasoconstrictor, though it is normally present only in exceedingly small quantities. Other molecules that mediate certain regulatory pathways may become unnecessary to transport after the vasculoid is installed, as for example the renin-angiotensin pathway (since there is no blood pressure to regulate). The capacity demand totals ~42 gm, which would require ~43 trillion tankers for full encapsulation. However, the RDA for all minerals, electrolytes, vitamins and trace elements is just 6.8-16.1 gm/day for healthy adults, a maximum of 1.87×10^{-7} kg/sec, requiring only 0.03 trillion tankers assuming ciliary transport parameters as described above. (Sodium and potassium ion transport in individual neurons is $\sim 10^{6}$ ions/discharge; assuming $\sim 10^{10}$ active neurons firing at near-maximum ~100 Hz, whole-brain ion transport rate is $\sim 10^{18}$ ions/sec or $\sim 7 \times 10^{-8}$ kg/sec, well within the aforementioned mass flow budget for mineral and electrolyte transport. However, Na^{+} and K^{+} are very quickly recycled after each discharge, so the true transport requirement is more accurately measured by the small amounts of these electrolytes that are lost in the gut, sweat, and lymph.) Transport of lactic acid, RNA/DNA, enzymes (e.g., those present as a result of normal tissue breakdown such as CPK and transaminases), hormones, cytokines and so forth (~1.8 gm) requires 1.9 trillion additional tankers. In some cases these "other" molecules may be shipped mixed in a single container.

A total of 166.2 trillion tankers are required to transport all-important physiological molecules at maximum human metabolic rates. At basal metabolic rates, only 11.78 trillion tankers will typically be active. If, contrary to our assumption, RDA levels are found to be an inadequate substitute for natural blood storage of plasma proteins, lipids, minerals and vitamins, or if production and demand occur at different times and the blood volume performs a significant storage function, then vasculoid design may need to include up to ~300 cm^3 of auxiliary bulk storage or caching to extend the "just in time" inventorying of these vital substances.

In addition to conveying molecules, the vasculoid must also transport physiologically important cellular species throughout the body. Cells to be transported (comprised mainly of white cells, platelets, and the few remaining red cells) generally range in size from 2-20 microns. These are shipped in cylindrical mostly-sapphire "boxcars" 100 microns long and 6 microns in diameter. These containers, of which 75 per cent (2120 $micron^3$) is usable storage volume ($\sim 2.4 \times 10^{-12}$ kg dry mass per boxcar), will just fit through average sized capillaries (~8 microns diameter, ~1000 microns in length) and all larger vessels, after accounting for plate thickness, and thus the boxcars have access to the vast majority of human tissues. The largest transportable cells (20-micron monocytes) must be temporarily dehydrated 50 per cent by volume prior to loading; such desiccated cells quickly rehydrate by natural osmosis upon release into the aqueous intercellular environment. Spherical cells 17.5 microns in diameter or smaller need not be dehydrated during loading, as they can fit into the available volume via change of shape. Typical transit time from point of entry to point of debarkation is ~70 sec (i.e., ~half the mean circuit length of 1.4 m, at 1 cm/sec), well within cellular ischemic survival times except possibly for oxygen, which may need slight supplementation during transit. Onboard computers and sensors embedded in interior boxcar walls allow the detection and interpretation of surface antigens on passenger cells to ensure that all transported cells are non-hostile, and if not, to refuse transport, convey them to a disposal site, or destroy them after entry. The following is an assessment of the minimum human cell transport requirement for the vasculoid appliance.

Under normal physiological conditions, 27-54 billion leukocytes circulate in the human blood volume, although in cases of chronic myelogenous leukemia the total count may increase to 0.13-2.7 trillion white cells. These 54 billion leukocytes consist approximately of 35-41 billion neutrophils (10-12 microns in diameter), 11-14 billion lymphocytes (8 microns), 1.6-4.3 billion monocytes (12-20 microns, average 15 microns), 1.1-2.7 billion eosinophils (12 microns), and 0.3 billion basophils (10 microns). Based on maximum container packing densities for each type of cell, 12-28 billion boxcars are required to encapsulate and transport all normal populations of bloodstream leukocytes. This will frequently include more than one cell per boxcar—multiple cells should be sedated for transport (using reversibly rotored molecular agents) to avoid unwanted interactions such as clonal expansion or induction of apoptosis. Principal origination sites are the bone marrow, the spleen, and the lymph nodes, but leukocytes may be distributed to almost any location in the body. Giant macrophages (e.g. >30 microns) and fibroblasts already present in the tissues (that normally do not enter the bloodstream) cannot be relocated by the appliance, although fresh stem cell precursors of fibroblasts with appropriate activating cytokines can be relocated from their natural source site, as required. A comparatively small population of circulating stem cells must also be transported by the vasculoid. For example, vasculogenesis is assisted by hematopoietic stem cells (surface marker CD34+) that originate in the bone marrow and enter the peripheral blood in response to colony stimulating factors, with 1-3 per cent expressing the common leukocyte surface antigen CD45. Upon reaching the vascular surface and in the presence of the cytokine IL-3, these cells may differentiate either into leukocyte precursor (CD34+/CD45+) or into endothelial precursor (CD34+/CD45–); with the cytokine VEGF present, the latter cells further differentiate into new endothelial cells. The vasculoid must provide transport for stem cells of these and other types, along with the required cytokines, to ensure vascular repair of damaged endothelium and for other purposes. Fortunately there are only a limited number of such cell types, and the total population of stem cells requiring transport is many orders of magnitude smaller than the population of leukocytes. As a result, stem cell transport should be a comparatively minor—if physiologically essential—task for the previously-specified boxcar fleet.

Although substitution of vasculoid transport for liquid blood eliminates much of the conventional health risk due to bleeding, blood vessel breaches that penetrate the endothelium but do not penetrate the vasculoid will still exude large amounts of extracellular serous fluid and other critical substances if the leakage is not promptly staunched. Furthermore, platelets are storehouses for a variety of molecules that affect vascular tone, fibrinolysis (subsequent clot dissolution), and wound healing. These substances are released during the clotting process. Platelets also communicate via chemical messengers with other blood cells (e.g. macrophages and fibroblasts) and with the endothelial cells that coat the interior of all blood vessels. Hence, platelets still have an important role in human physiology even after the vasculoid has been installed. Approximately 2 trillion platelets (~2 microns in diameter) circulate in human blood. Each boxcar can hold about 500 platelets, giving a requirement of 4 billion additional containers for platelet transport. As with white cells, multiple cells could easily activate and aggregate, hence their function should be heavily suppressed during transport (again, using reversibly rotored molecular agents).

In the vasculoid, all respiratory gases are transported in bulk and at high pressure. Boxcars could in principle be used to transport red cells, but RBCs have an effective storage pressure <<1 atm which is far less efficient than the 1000 atm storage pressure of the respiratory gas tankers. Consequently, the ~30 trillion red cells normally present in the human bloodstream are superfluous and may be permanently removed from circulation. Beyond their respiratory function, red cells also are mechanically involved in the clotting process. However, the vascular plug typically contains mostly platelets, plasma fibronectin and factor XIII-crosslinked fibrin, plus small amounts of tenascin, thrombospondin, and SPARK (secreted protein acidic and rich in cysteine) with a relatively minor RBC contribution, so RBCs probably are not essential to the clotting process or are necessary only in very small numbers. A sufficient oxygen concentration maintained in kidney peritubular cells should reduce erythropoiesis to at most 1 per cent of the normal rate of red cell production , requiring only ~34,700 red cells/sec to be conveyed from the erythroid marrow directly to the liver or spleen for disposal. Alternatively, erythropoiesis can be reduced by directly regulating the amount of erythropoietin that is allowed to reach the bone marrow. With few erythrocytes remaining to be destroyed in the liver, hepatic heme catabolism drops significantly, greatly reducing the production of bilirubin, a yellow bile waste pigment that plays no role in fat digestion in the gut but imparts the brown color to feces; as with hepatitis patients, the stool of the envasculoided user may become chalky white. Ciliary transport of boxcars containing the few remaining red cells over a ~1.4 meter circuit at a conservative mean velocity of 0.1 cm/sec (allowing plenty of time for docking and cell unloading) requires a circulating subfleet of just 34.7 million boxcars (only ~0.1 per cent of the entire boxcar fleet).

A maximum of 32 billion boxcars are required to convey all essential bloodborne cells during normal physiological conditions, although at times as few as 16 billion boxcars may be actively involved in cellular transport. Since a principal benefit of vasculoid installation is a significant decline in susceptibility to microbial invasion, the above figures can probably be further reduced in actual clinical practice. The interior volume of the vasculoid is lined with a sapphire surface (comprised of watertight adjacent installed plates) from which protrude trillions of mechanical cilia (arranged in patterns designed to maximize transport speed and reliability. A secondary role is to assist the vasculocytes (small mobile nanorobots) in cleaning up after accidental internal spills, component malfunctions, or other pathological events. Even by 1997, prototype MEMS ciliary arrays of up to 1024 cilia had already demonstrated transport speeds up to 200 microns/sec with ~3 micron positional accuracy. While a highly efficient specialized ciliary mechanism can undoubtedly be designed, for the present study each vasculoid cilium is conservatively assumed to be similar in size, shape, and performance characteristics to the vacuum-sealed robotic manipulator arm described by Drexler. In brief, each cilium is a cylindrical assembly 100 nm long and 30 nm in diameter, having a transverse travel of 100 nm and a lateral speed of 1 cm/sec, consuming 0.1 picowatt during continuous operation at low load. Note that even if the vasculoid interior contained gas at 1 atm, viscous drag would contribute an additional power loss of only ~0.001 pW per cilium (taking viscosity h_{air} = 1.83 x 10^{-5} kg/m-sec at 293 K). Mean cilium/cargo contact time is ~10 microseconds. This implies an acoustic frequency >100 KHz, well above the conventional human threshold of hearing; however, ultrasonic hearing up to 108 KHz via otolithic conduction has been reported in humans, so further study is required to ensure restriction of transport motions to sufficiently high frequencies to preclude any perceptible hum directly due to ciliary operations, and possibly involving a variety of noise cancellation, anti-resonance (e.g., out-of-phase operation) and acoustic dampening techniques. Additionally, physical discomfort, intra-articular pain, and a lowering of electrical pain sensation threshold due to ultrasound exposure have also been reported in humans. The possibility of indirect audible or infrasonic vibrations due to beat frequencies, spatial acoustic interference patterns, or various resonances should be investigated further, along with the possibility of undesired vibrational energy losses into the tissues. Each cilium can apply ~1 nanonewton at the tip. This is sufficient force to accelerate a 1 $micron^3$ block of water (weight ~0.01 piconewton) at 10^5 G, or a 100-micron long, 6-micron wide water-filled sapphire-shell cylindrical boxcar at ~30 G. The human body is normally subject to accelerations far less than 30 G; with tenfold grasping redundancy, the probability that a container will be shaken loose from the cilia and tumble uncontrolled into the vascular lumen, due to normal macroscale accelerations of the patient's body, is extremely remote. In the rare instance where such detachment occurs, the container will eventually be caught and returned to the flow; kinetic impact energy in such cases is relatively small. Whether such cross-luminal impacts can trigger a detachment avalanche should be investigated using computer simulations. Changeable cilium tool tips may permit rapid, reversible, "blind" container grappling using van der Waals adhesive forces, which forces may be controlled by varying adhesive pad surface corrugations at the nanometer scale, or by other means. The cilia might also have a reversing function (e.g., for loosening container jams). Molecular dynamics simulations of these processes would be useful.

The principal task of the ciliary distribution subsystem is to reliably transport a maximum of 166.2 trillion 1-micron cubical tankers and a maximum of 32 billion cylindrical 6 mm x 100 mm boxcars through a physical circuit ~1.4 meters in length. Because of their smaller size and greater number, ciliary subsystem design is driven almost entirely by tanker, not boxcar, requirements. Although a single cilium can apply sufficient force to grapple and manipulate containers of either size, tenfold grappling is employed in the baseline design to achieve firm contact, to avoid physical escape of containers from the cargo traffic stream, to ensure massive redundancy hence high subsystem reliability, and to allow extremely precise steering. If a tanker occasionally encounters a patch of 9-contact cilia, or the even rarer 8-contact patch, steering or grappling capability should not be significantly reduced. Fail-safe grappling modes must be designed for cilium end-effectors. Each tanker is normally in continuous contact with 10 cilia at all times during transport, so maximum ciliary spacing is 0.32 micron. A minimum of 3000 trillion cilia are needed to achieve 100 per cent service over the entire ~300 m^2 vascular (mostly capillary) surface. A set of 3000 trillion cilia physically occupies ~2.1 m^2 (~7 per cent) of the vascular surface. Assuming natural background radiation damage causes 1.5×10^7 fatal hits/kg/sec for nonredundant structures, the mean time to failure of a ~10^{-19} kg cilium is ~10^{12} sec, which implies a mean failure rate of only ~2000 cilia/sec (~20 nanograms/day) throughout the entire vasculoid. Using redundant structures mean failure rate may be considerably reduced. Under resting metabolic conditions, 11.78 trillion tankers are transported by 117.8 trillion cilia generating waste heat of ~11.8 watts, an increment of only ~12 per cent over the basal human metabolic rate. During periods of maximum exertion and thermoregulatory stress, 166.2 trillion tankers are transported by 1662 trillion cilia generating excess heat of ~166 watts, well below the maximum human metabolic rate of 1600 watts and just slightly exceeding the ~100 watt effective load error that could trigger a response from the human thermoregulatory control system.

Once molecules or cells have been transported to the appropriate site, the container in which they reside must dock with the vasculoid surface on the interior side, and pass the delivered materials through this surface to the interstitial side of the vasculoid wall, whereupon the cargo can diffuse or migrate into the tissues as required. Tankers offload their molecular cargo in docking bays whereas boxcars unload their cellular passengers at cellulocks. Possible undesired long-term chemical interactions of transported nutrients discharged into the fluids of the vasculoid-endothelial interface at above-physiological concentrations must be carefully analyzed. For example, high concentrations of glucose akin to chronic hyperglycemia could enhance protein glycosylation of long-lived proteins (e.g., vascular and myocardial collagen) which undergo continual cross-linking during aging because of the formation of advanced glycosylation end-products (AGEs) , most significantly (for this paper) in endothelial cells , which may be partly responsible for arterial stiffening and higher blood pressure with age (and which drugs like ALT-711, benfotiamine , nitric oxide and other substances may reverse). AGEs can also induce apoptosis and vascular endothelial growth factor overproduction in capillary pericytes.

Physiologically relevant molecules are delivered to specific locations by docking at 2 $micron^2$ (~1.4-micron square) tanker docking bays embedded at appropriate spatial intervals across the vasculoid surface. After securely connecting to the docking bay structure, and after the tanker's cargo manifest is scanned by the docking bay computer and said cargo is approved for receipt (Section 5), the tanker can be pumped dry by molecular sorting rotors in ~10 seconds, then sealed, remanifested, undocked, and released empty back into the traffic. Empty tankers may be reloaded with molecular cargo by a similar process in a similar time. Note that tankers carrying liquid or gas may discharge their contents into the narrow end of a funnel-shaped manifold leading to the active rotor banks at the wider end, to minimize the required tanker connection aperture. In different operating modes, tankers might be only partially emptied at each stop, or might be partially emptied more rapidly via bulk-flow exhausting rather than molecular rotoring. Docking bays include buffer tanks so that offloaded materials can be metered out to the underlying tissues over time periods longer than 10 seconds. If the average endothelial cell (of which blood vessel walls are comprised) has a flat luminal area of 300-1200 $micron^2$, then the ~300 m^2 vascular surface may be comprised of up to ~1 trillion endothelial cells. This vascular surface could theoretically accommodate a maximum of 300 trillion tankers positioned side by side. At basal metabolic rates, only 11.78 trillion tankers are actively on the move. Even during peak exertion, a maximum of 166.2 trillion tankers are being transported, requiring just 55 per cent of the available vasculoid luminal surface for their passage assuming all tankers are traveling in monolayer. Other functionally equivalent modes might include operating all tankers at reduced capacity or reduced duty cycle during periods of basal demand, or reducing the transport velocity. A detailed computer simulation study of optimal traffic patterns near peak capacity would be valuable to perform. A maximum of 166.2 trillion tankers circulating through an average ~1.4 meter circuit at ~1 cm/sec implies a docking bay cargo unloading requirement of ~1.2 trillion tankers/sec. Docks operate on a 50 per cent duty cycle to allow plenty of time for maintenance and repair, although in a less conservative design the duty cycle could possibly be boosted as high as 99 per cent. Tankers are emptied in 10 sec, so ~24 trillion docking bays are needed for reliable operation at peak loads. Docking bays occupy ~16 per cent of the vasculoid surface and have a mean center-to-center separation of ~3.5 microns. Each endothelial cell is serviced by up to ~24 docking bays. Normally only a small fraction of the body's tissue cells are directly bathed in blood. Hence, delivery of molecules to each endothelial cell surface must be sufficient to supply the needs of the up to ~30 tissue cells that lie, on average, within ~3 cell widths of the nearest capillary. The maximum whole-body transport requirement of 2.25×10^{22} molecules/sec, consisting principally of respiratory gases, water, and glucose, gives an endothelial surface mass transport rate of 2.25×10^{10} molecules/sec/endothelial cell. This transport rate may be satisfied using a minimum of 45,000 molecular sorting rotors per endothelial cell operating on a 50 per cent duty cycle. With 24 docking bays per endothelial cell, the requirement is ~0.94 billion molecules/sec per docking bay. Allowing tenfold multiplicity of sorting rotors to ensure high reliability requires 18,750 sorting rotors per docking bay, of which only 67 will be active at the basal rate and up to 9375 will be active during maximum physical exertion; 18,750 sorting rotors measuring 98 nm^2 on the ventral surface (facing the endothelium) cover 1.84 $micron^2$, essentially coating the entire undercarriage of the 2 $micron^2$ docking bay facility. This provides a ~7.5×10^8 molecule/sec supply per tissue cell.

At peak metabolic load, maximum molecule offloading rates as constrained by possible effervescence and crystallescence effects appear acceptable in this application for all low-volume molecules and for glucose and CO_2, but unfortunately exceed the effervescence limit for O_2 by a factor of ten, due to the relatively poor aqueous solubility of oxygen. Hence at peak load this gas must be offloaded either ~10 times more slowly than indicated above for a single release site, or at the indicated flow rate from ten discrete release sites all spatially separated by more than the largest possible O_2 effervescence bubble radius. To simplify rotor system design, there are four classes of docking bays approximately matching the anticipated four principal tanker populations. Of the ~24 docking bays overlying each endothelial cell, ~5 stations are designated to handle tankers carrying respiratory gases, and possess rotors (on the tissue side) designed to reversibly bind only oxygen or carbon dioxide molecules. Another ~14 docking bays receive only water-bearing tankers, and employ rotors specializing in water molecule transport. Another ~3 stations accept only glucose cargoes, and use only glucose sorting rotors. The remaining ~2 docking bays are general-purpose stations equipped with rotors of up to tens of thousands of different types capable of reversibly binding all remaining biologically useful molecules (or classes of molecules) that must be transported through the vasculoid surface.

In addition to gas, water, glucose, and specialty tankers, there is a fifth class called power tankers that support the docking bay oxyglucose metabolism. Power tankers transport stoichiometric parcels of oxygen and glucose, and also remove carbon dioxide byproduct for disposal at the lungs. (Vasculoid power may be generated by glucose engines, glucose fuel cells, or other energy conversion devices that consume oxygen and native glucose; water produced by glucose engines may be vented into the biological tissues) All docking bays can accept power tankers, which supply the docking bay's own internal energy needs. Tankers reload their own oxyglucose engines and fuel tanks as they make their rounds of the various docking bays. Tanker duties are not energy intensive, so tankers refuel at most once a day, at the docking bays. Rotors conveying molecules from tankers (at high concentration) through the vasculoid integument to the endothelial surface (at low concentration) generate positive mechanical energy, which may be used to largely offset the work of concentration that must be expended to transfer molecules in the reverse direction when loading tankers from the tissue space. Concentration work losses can be reduced almost to zero in an efficient system, with systemwide entropic losses in the compression-expansion cycle as low as 0.01-0.2 watts. Hence the primary energy loss is the drag power per rotor, ~10^{-16} watts or ~0.1 zJ/molecule assuming a transfer rate of ~10^6 molecules/rotor-sec. Taking the whole-body transport requirement range as 10-225 x 10^{20} molecules/sec, the net molecular transport power dissipated per docking bay is thus 0.004-0.09 picowatt depending on physical exertion level, or 0.1-2.25 watts for the entire vasculoid docking bay subsystem. (Rotor systems that do not include efficient energy recovery and must dissipate the work of concentration, perhaps up to ~20 zJ/molecule, may require different appliance architecture) To hold docking bay subsystem total power requirement for computational tasks to a 10-watt budget, nanocomputers using an appropriate mix of local and distributed computing costing ~6 x 10^{-17} watts/(ops/sec) can provide ~200 billion MIPS systemwide or ~10,000 ops/sec per docking bay, costing ~0.6 pW per docking bay. These rates have been enhanced by employing reversible logic. Hence the total power per docking bay for both mechanical and computational tasks ranges from ~0.604 pW (basal) to ~0.69 pW (peak), or 14.5-16.6 watts for the entire system of 24 trillion docking bays. If power is supplied by a glucose engine achieving mechanochemical power conversion at ~10^9 watts/m^3, a 16.6 watt peak energy budget for all docking bays requires one glucose engine only ~0.0007 $micron^3$ in volume (~0.03 per cent of bay volume), per dock, and a similar volume of fuel tankage. Biological cells are delivered to specific locations by docking at cellulock stations embedded at appropriate intervals across the vasculoid surface, often preferentially located near specialized post-capillary blood vessels less than 30 microns in diameter found in lymphoid tissues, commonly known as high endothelial venules (HEVs). Cellulock density requirements also vary on a tissue-by-tissue basis—for instance, intestines, lungs, throat and nasal cavity probably have much higher requirements than heart or brain. After securely docking with their loading face pressed firmly against the vasculoid surface, boxcar doors dilate open to make a 30 $micron^2$ (~6.2-micron diameter circle) aperture. Cellulock doors (constructed of self-cleaning sliding spiral graphene segments in snug facial contact) then dilate to the same diameter, exposing passenger cells to the interstitial fluid. Passenger cells are gently pistoned out through the opening, taking care to minimize extracellular fluid entry. The cycle ends as all doors reseal and the boxcar undocks, resuming its travels. Biochemical signals of cellular distress (e.g. cytokines such as interleukin-1 or tumor necrosis factor) are received by sensors on the interstitial side of the vasculoid. Upon receipt of this chemical signal, the cilia at that site are programed to call for leukocyte activity there. Cellulock and boxcar interiors may employ a small set of presentation semaphores which expose to the interstitial fluid minute quantities of various chemoattractants and chemorepellents keyed to each major type of cell to be transported. These chemicals are permanently bound to the rotors and are not released from the vasculoid; like semaphores, each rotor position may present differing concentrations of one chemical species, or a selection of different chemical species. White cells of the type desired to be transported can be encouraged to enter or exit, as required, by positioning presentation rotor settings so as to create "chemotactic funnels"—programable patterns of varying concentration gradient to steer motile cells toward, or away from, the cellulock aperture (e.g., haptotaxis). Contact sensors around the aperture detect the presence of a motile cell "requesting entry" to the transport system. Motile cells presenting themselves for entry at the interstitial side of the cellulock are, at minimum, subjected to a rudimentary cell protein coat identification test before they are allowed admittance. Affirmatively recognized harmful foreign pathogens such as bacteria or viruses and infected native motile cells may be transported to a specific disposal site or destroyed, rather than being admitted freely for transport elsewhere in the tissues. The cellulock mechanism covers 60 micron^2 (~7.7-micron square) of vasculoid surface. If spiral door segments slide past each other at 0.004 m/sec over an average of one-half rotation or ~15.4 microns, dilation (or resealing) of the cellulock door takes 3.85 milliseconds. Sliding friction for 16 nm^2 contact surfaces traveling 20 nm at 0.004 m/sec dissipates ~10^{-24} joules, or ~3.13 joules/m^2-meter . Thus ~30 micron^2 cellulock sliding door surfaces moving ~15.4 microns to open, then again to close, require ~3 x 10^{-15} joules to complete one dilation cycle. Assuming a maximum rate of 32 billion boxcars per (normal blood circulation time of) ~60 sec to emulate typical physiological delivery rates, the population of vasculoid cellulocks must process a total of 533 million boxcars/sec. If there are therefore also 533 million cellulock dilation cycles/sec, frictional waste heat generated by moving door segments in the entire vasculoid cellulock subsystem is a negligible ~1.5 microwatts. Taking cellulock power as equal to the basal rate for 30 docking bays gives a requirement of ~18 pW/cellulock, or ~0.59 watts for the entire cellulock subsystem. Serum glucose and oxygen to provide cellulock power (a minor volumetric requirement) may be rotored in directly from the interstitial fluid on the external side of the vasculoid surface. Since the active tanker/boxcar ratio varies from 736 (basal) to 5194 (max) depending on metabolic load, there should be no less than one cellulock per 736 tanker docking bays, or a total of 32.6 billion cellulocks in the vasculoid surface, again allowing a 50 per cent duty cycle for maintenance and repair. Thus there is approximately one cellulock for every boxcar in circulation. Cellulocks occupy ~0.65 per cent of the vasculoid surface and have a mean center-to-center separation of ~96 microns. Each cellulock services an average of ~31 endothelial cells, though the actual spatial distribution of cellulocks is expected to be highly nonuniform (e.g., clustering near HEVs). The average capillary is ~1 mm long and is made up of ~60 endothelial cells, and there are ~19 billion capillaries in the human body, so on average ~1 cellulock resides at either end of every capillary vessel in the body. The maximum white cell transit rate through injured or diseased tissue is ~1 cell/sec/mm^3. With 32.6 billion cellulocks there are ~163 cellulocks/mm^3 of tissue. Each cellulock has ~163 sec to pass each white cell, even at maximum traffic flows. This should be sufficient, given that cellulock doors open and close in milliseconds, and natural white cell transendothelial diapedesis may occur in as little as ~3 minutes. The creation of unnecessary interior voids where unwanted opportunistic pathogens might flourish should be avoided. It is probably unnecessary to maintain voids between vasculoid and endothelium near cellulocks because: (1) the time allocated to cycle the cellulocks appears adequate to permit WBC diapedesis, and (2) diapedesis can probably be expedited using chemical factors emitted by the vasculoid.

The ideal method for container routing will be completely automatic, consuming no incremental power or compute cycles. To this end, the outer surfaces of each of the five types of tankers (Gas, Water, Glucose, Power, and Other) and the four types of boxcars (Red Cell, White Cell, Platelet, and Other) bear a specific repeating interlock pattern, crudely analogous to Braille. These patterns remain in place on the container surface unless the container is affirmatively "reset" to carry a different cargo, in response to varying physiological needs and local container availability. Cilia surrounding each docking bay or cellulock are outfitted with gripper pads that adhere adequately to all container surfaces, but strongly only to the surface pattern of the container type sought by the docking bay or cellulock with which the cilia are associated. These pad "recognition" types normally remain constant over time, but, like the container surfaces, may be modified as required in special circumstances. Efficient methods for handling other-type tankers that can avoid the need for time-consuming docking and downloading of manifests should be investigated further. Cilia that have "recognized" a desired container transmit a trigger signal to nearby cilia, activating a collective stereotypical docking sequence. Thus, although containers follow statistically random paths across the vasculoid surface, each docking bay and cellulock automatically receives the cargo it requires from the passing container stream. Docking acquisition is demand-regulated locally.

The exemplar vasculoid has been scaled to transport essential molecules and cells at maximum physiological transfer rates. However, all engineering systems involve design tradeoffs and limitations, and the present design invokes a number of potential transport bottlenecks including intestinal water absorption and renal water elimination, glucose absorption and elimination at peak loading, boxcar geometrical clearance in narrow capillaries, and geometrical limitations of tanker monolayer transport.

Water Absorption and Elimination—The ~6 meter-long intestinal tract has a simple cylindrical surface area of ~0.65 m^2, but the ~5 × 10^6 villi of the small intestines (which is most of the length) increase the effective absorbing surface to ~10 m^2. The ileum, the lower portion of the small intestine, absorbs water into the bloodstream at the maximum rate of ~0.07 cm^3/sec. The cylindrical villi, typically measuring ~200 microns wide and ~1000 microns long, are heavily capillarized; assuming ~1000 capillaries/mm^3, each villus has ~30 capillaries presenting a total ~8 × 10^5 $micron^2$ of absorbing surface—for the entire ileum, that's ~4 × 10^{12} $micron^2$, a water transport rate of ~6 × 10^8 molecules/sec-$micron^2$. If maintenance activities are temporarily suspended during peak demand, water tankers loaded with 2.51 × 10^{10} water molecules at 2 $micron^2$ docking bays can provide the same transport rate of 6 x 10^8 molecules/sec-$micron^2$ if 49 per cent of the vasculoid surface in the ileum is given over to water transport—potentially crowding out some "applications plates" at this surface. (The requirement for alimentary water can also be greatly lessened by reducing the need for water solvent in urine by reducing the need for water cooling in sweat and possibly by reducing water losses in the alveoli (which could be covered with an improved waterproofing surfactant.) Excretion of ~0.023 cm^3/sec aqueous waste at the kidneys occurs primarily through the renal glomeruli which have a total surface area of ~1.4 m^2, giving a maximum water export rate of ~5 × 10^8 molecules/sec-micron^2. If maintenance activities are temporarily suspended during peak demand, water tankers loaded with 2.51 × 10^{10} water molecules at 2 micron^2 docking bays can provide the required transport rate if 40 per cent of the vasculoid surface in the glomeruli is given over to water transport. The ~0.001 gm/sec (~4 x 10^6 molecules/sec-micron^2) solute waste stream is conveyed on mixed-cargo tankers carrying ~10^9 solute molecules per tanker, which may be unloaded at mixed-cargo docking bays covering only 8 per cent of the vasculoid surface in the glomeruli.

Glucose Absorption and Elimination—Glucose is absorbed in the small intestine, mostly in the duodenum and the jejunum, at a maximum rate of ~10^{20} glucose molecules/sec, somewhat less than our maximum glucose requirement of 5 x 10^{20} molecule/sec. Given a capillary absorptive area of ~3 x 10^{12} micron^2 in the upper small intestine, the natural glucose transport rate is ~3 x 10^7 molecules/sec-micron^2. As before, if maintenance activities are temporarily suspended during peak demand, glucose tankers loaded with 3.93 x 10^9 glucose molecules at 2 micron^2 docking bays can provide the same transport rate if 17 per cent of the vasculoid surface in the duodenum and jejunum is given over to glucose transport. A related issue is the vasculoid's response to severe gastrointestinal overloading, as for example the rapid ingestion of a large quantity of pure sugar by the user. Sugar has a tremendous irritating effect on the stomach, with 15-30 gm (60-120 kcal) producing great outpourings of stomach mucous; as little as 0.25 liter of 20 per cent sugar solution (78 gm, 300 kcal) is sufficient to cause vomiting in some patients with chronic gastritis, according to century-old laboratory experiments. A rapid 2000 kcal (1.75 x 10^{24} glucose molecules) ingestion would represent a bulging stomach-full (<1.5 liters) of >22 per cent sugar solution (~526 gm of sugar) which would be severely irritating to tissues and toxic to cells, almost certainly exceeds the natural absorption abilities of the small intestine, and would likely prove emetic. Urination at the physiological maximum ~10^{20} glucose molecule/sec elimination rate unloads a 2000 kcal sugar binge in ~5 hours, a renal export rate of 7.14 × 10^7 molecules/sec-micron^2 that can be handled at glucose docking bays covering 36 per cent of the vasculoid surface in the renal glomeruli. Alternatively, some excess sugar can be routed to the liver or other organs for conversion to glycogen or starch.

Boxcar Clearance in Narrow Capillaries—Cell-carrying boxcars 6 microns in diameter cannot squeeze through the narrowest of capillaries, most notably those found in the human retina which may be as small as 4 microns wide. This has two operational implications. First, boxcars carrying cellular cargo destined for tissues having capillaries <8 microns must onload and offload at cellulocks located outside such tissues, relying on natural transtissue mobility to achieve the desired placement of the transported cells. In this the vasculoid mimics the natural transport process. Second, boxcars must be capable of being rerouted around narrow capillary bottlenecks to their flow, else the carriers may lodge in the narrow passage and become an obstacle to the free flow of materials as sometimes occurs with leukocyte plugging in certain pathological conditions such as diabetes. Cilia with boxcar-coded gripper pads located at the entrances of difficult passages may be programed to refuse boxcars, and may also utilize reverse ciliary flows to assist this process. Alternatively, boxcars may be designed with flexible metamorphic surfaces (and thus a limited ability to change shape) to allow tight passage when traveling empty. With metamorphic boxcars, cilia can allow boxcars to pass that will empty their contents before reaching a capillary. Another alternative is that the vasculoid could be used directly to induce changes in capillary size.

Tanker Monolayer Transport—If container traffic is strictly limited to a single monolayer at the artery or vein wall, then ciliary transport produces a tanker bottleneck in medium- and large-diameter vessels of the arteriovenous (noncapillary) vasculature. This is most clearly illustrated in the aorta. The circulation time of ~140 sec (1.4 m circuit at v_{tanker} = 1 cm/sec transport speed) implies a system flux of 8.4 x 10^{10} tankers/sec at basal load and 1.2 x 10^{12} tankers/sec at peak load. However, the luminal surface of an aorta of diameter $2R_{aorta}$ = 2.5 cm has room to accommodate only a flux of 2p R_{aorta} v_{tanker}/S_{tanker} = 7.9 x 10^8 tankers/sec per monolayer of coverage, assuming S_{tanker} = 1 mm^2 tankers traveling at v_{tanker} = 1 cm/sec. The operational implication is that most tankers must be allowed to transit the aorta as a multilayer averaging 0.11 mm deep (~110 tanker layers) or <1 per cent of available luminal diameter at basal load, and up to 1.5 mm deep (~1,500 tanker layers) or ~12 per cent of available luminal diameter at peak load. This bottleneck extends to a lesser degree throughout the arteriovenous vasculature, with the number of required tanker transport layers decreasing rapidly with decreasing vessel diameter. Purely monolayer tanker transport becomes tenable in blood vessels with diameters <200 microns (e.g., arterioles and large venules) at basal load or <20 microns (e.g., metarterioles) at peak load. Tankers can employ one or more simple grapples or temporary coupling mechanisms to join up into progressively lengthening chains (of linked tankers) as they pass blood vessel bifurcations moving upstream, and then to gradually detach and shorten the chains as they move downstream. Even a single cilium can exert sufficient force to pull a fully-loaded 1000-tanker chain at ~100 G, although the tanker at the base of each chain will normally be grasped by up to 10 independent cilia. Nevertheless, chain linkage reliability and the effects of such extended chains on flow viscosity should be studied more thoroughly. Other approaches, including facultative lateral links between chains to improve transport stability, tanker bloc motions, independently ciliated tankers (allowing high flow rates in large vessels with low inter-laminar slip rates), mechanically valved free flow, extensive water and respiratory gas caching to reduce tanker count, and increasing transport speeds up to 1 m/sec in large blood vessels, are potentially useful strategies but will not be explored further here.

An additional small population of mobile legged nanorobots, or "vasculocytes," continuously patrols the vasculoid interior. Vasculocytes are independent nanodevices several microns in size, replete with ambulatory appendages, manipulator arms, repair and assembly tools/materials, onboard computers with mass memories, communications and sensory equipment, and independent power supplies and fuel tankage. A description of a similar system is in; a fuller discussion of these capabilities is in and is beyond the scope of this paper. Like other vasculoid subsystems, to ensure high reliability vasculocytes are designed with the usual tenfold redundancy in most major components. Vasculocytes in the vasculoid appliance have many important functions, including: (1) general maintenance and repair activities; (2) plugging internal leaks and breaches, cleaning spills, leak scavenging, and maintaining an orderly particle-free internal environment; (3) repairing or replacing malfunctioning ciliary leg mechanisms (designed for modular changeout); (4) clearing of jammed docking bays or cellulocks; (5) reconstruction of the vasculoid sapphire surface plate array after a breach or other catastrophic failure; (6) physical reconfiguration and remodeling of the vasculoid microstructure, as for example to extend additional segments into new capillaries following angiogenesis, to remove vasculoid tendrils from dead capillaries, or to expand or contract existing segments (details deferred to another paper); and (7) installation and removal of vasculoid systems from the human body. With a vasculocyte mass of ~10 picograms, volume ~3 micron^3, and peak power consumption ~50 picowatts to allow bursts of more intensive computation (at up to nearly ~1 megaflop/sec) or other activity, a 10-watt vasculocyte subsystem energy budget permits the deployment of 200 billion continuously active devices, a total 2 gm mass of nanorobots. Deployment of this number of devices would allow, for example, at least one vasculocyte to visit all 24 trillion docking bays once a day, spending 720 sec visit, enough time to perform ~10^9 manipulatory operations using a 10^6 Hz robotic arm . Alternatively, this would allow 7200 visits/day per docking bay, each visit lasting ~100 msec (~10^5 operations; travel time between adjacent bays at 1 cm/sec is only 0.35 msec) which is long enough to run a simple diagnostic set and to detect plate malfunction within just ~12 seconds (on average) of its initial occurrence. It would also allow one visit per day to each of 3000 trillion cilia, spending ~6 sec/visit (~10^7 operations). This seems sufficient. Assuming ~0.1 pW per vasculocyte leg (similar to cilium) and 4 legs in motion, simple walking consumes only ~0.4 pW of mechanical power; a 10,000 ops/sec computation budget to control simple walking, costing ~6 x 10^{-17} watts/(ops/sec), adds another ~0.6 pW, giving a total power requirement of just ~1 pW for continuous simple walking. An onboard 1 micron^3 fuel tank with energy storage density of ~2 x 10^{10} J/m^3 for externally-supplied O_2 gives each 1-50 pW vasculocyte a range of 400-20,000 sec between refueling stops, or ~100-5,000 sec if the O_2 is stored internally. The power draw of the entire 200 billion vasculocyte fleet is 0.2-10 watts. During vasculoid installation, the patient is injected with a population of vasculocytes adequate to construct the initial configuration in a reasonable period of time. After installation, a sufficient number of spare vasculocytes are stored internally in dormant condition in the vesicles until needed for repair, maintenance, or injury response functions. Active vasculocytes have highly redundant subsystems, hence fail infrequently. Upon such infrequent failure, they are replaced with reactivated dormant devices (self-repairing vasculocytes are possible in principle—e.g., modular repair-by-replacement—but would add significantly to overall system complexity, hence are not included in this scaling study.). Nanorobot detritus and internal waste is containerized and ciliated to an appropriate internal holding area located in the vesicles. New vasculocytes, replacement modules, and other necessary spare parts and repair materials may be exchanged into the user as consumables, when required. Vasculocytes performing general repair and maintenance on tanker docking bays have been allocated up to 10 sec docking cycle for this task a very conservative 50 per cent duty cycle for docking bays. Vasculocyte system subtasks which must be subsumed within this time budget may include: (1) Detection and analysis of system fault; (2) determination of appropriate response; (3) dispatch of appropriate repair mechanism to problem site, including travel time and docking; (4) on-site verification that the problem is as reported, and that the repair plan is correct; (5) deployment of tools and materials; (6) performance of repair work possibly including plate changeout, rotor replacements, sensor recalibration, etc.; (7) verification of correct completion of repairs and that the faulty subsystem is now performing properly; (8) retracting/repacking all tools with final verification of job completion; and (9) undocking and return to transport stream for storage or to receive instructions for the next assignment. The vasculoid configuration, including numbers of components, their locations, their accessibility for repair and their ease of disconnection/reconnection will influence the rapidity of these steps. Of course, most of the time there will be no fault at a particular docking bay, and running a diagnostic is much quicker than correcting a fault. Since there are only enough vasculocytes to allow up to 720 seconds/day of repair and maintenance activity at each docking bay, and since a 50 per cent duty cycle implies that each docking bay is allocated ~43,200 sec/day for this activity, then it might be possible to significantly increase the docking bay duty cycle up to (86,400—720) / (86,400) ~ 99 per cent in a less conservative design.

Power is supplied to the vasculoid appliance via the chemical combination of oxygen and glucose, both of which are plentiful in the human body. Total body power dissipation is ~100 watts at the basal rate and ~1600 watts at peak. As a further point of comparison, the basal human heart power output is $p_{heart} \sim P_{systolic} V_{blood} / t_{circ} = 1.4$ watts (taking pumping pressure $P_{systolic} = 120$ mmHg (1.59×10^4 N/m^2), pumped blood volume $V_{blood} = 5.4$ liters, and circulation time $t_{circ} = 60$ sec), rising to ~8.0 watts when cardiac output reaches ~30 liters/minute at peak exertion, assuming no rise in pressure. Experiments confirm that the basal power output of a 350 gm resting heart is ~3.5 milliwatts/gm or ~1.2 watts, and patients with chronic congestive heart failure can achieve measured peak cardiac outputs no higher than ~2 watts; efficient total artificial hearts (TAH) may draw ~5 watts of power. The four most significant subsystems requiring a regular energy supply include ciliary transport (11.8-166.2 watts), docking bays (14.5-16.6 watts), vasculocytes (0.2-10 watts), and cellulocks (~0.6 watts), giving a requirements range of 27.1 watts (basal) to 193.4 watts (peak). There are also 125 trillion applications plates which are unused in the basic vasculoid model described in this paper. If each applications plate is allotted a 0.6 pW energy budget comparable to the docking bays, and if all possible applications plates are simultaneously activated,, then the applications subsystem could require an additional 75 watts, boosting the total power draw of the vasculoid appliance to 102.1-268.4 watts, which is still only at the borderline of thermogenic significance.

Besides providing systemic materials transport throughout the human body, the water component of blood also serves an important thermoregulatory function. Removing most of the circulating water from the vasculature eliminates one of the means by which the body regulates its gross thermal conductivity. In addition, the extremely high thermal conductivity of diamond imposes the requirement for a vasculoid design using mostly sapphire, rather than diamond, construction materials, even though diamondoid materials have been more extensively discussed in connection with molecular nanotechnology and medical nanorobots and may still be employed in lesser quantities (e.g., as thin coatings) throughout the vasculoid structure. Aside from blackbody radiation, sweating, and behavioral thermoregulation (including respiratory cooling), the body regulates its temperature and offloads excess heat principally via two mechanisms: *First,* there is passive conduction. Heat travels by pure conduction through fat and muscle from the body core out to the periphery. The thermal conductivity of human tissue is $K_t \sim 0.5$ watts/m-K, so for a typical L = 10 cm path length (~half-torso thickness), heat flow $H_f \sim K_t/L = 5$ watts/m^2-K, or ~10 watts/K for a 2 m^2 human body. In a cold room, the mean temperature differential between core and periphery DT ~ 11 K, so $H_f \sim 100$ watts, which is approximately the basal metabolic rate. Experiments confirm that 5-9 watts/m^2-K is the minimum heat flow in very cold conditions (the actual value depending largely upon the thickness of subcutaneous fat layers). In this case, the peripheral capillary blood flow has slowed to a trickle, producing the minimum thermal conductivity of the human body in cold conditions. On the other hand, in a warm room or during heavy exercise, DT ~ 1 K, so $H_f \sim 10$ watts. Thus, paradoxically, at warmer temperatures when the human body is generating considerable surplus heat, the body's passive heat flow is actually very low because of the smaller temperature differential between core and periphery. *Second,* heat is transported via the active blood flow. In warm rooms, not only are the peripheral capillary sphincters fully dilated, allowing more blood to flow through the peripheral capillaries relative to the core capillaries, but also the total volume of blood flow may increase. (During heavy exercise, total blood flow volume may rise by a factor of 4 or 5.) Diathermy experiments suggest that the active blood flow mechanism alone may carry off 100-200 watts of heat before core temperature starts to rise. In cold rooms and in the absence of heavy exercise, peripheral capillary sphincters are maximally contracted, thus minimizing blood flow (and hence heat transport) to the periphery.

To summarize: The passive conduction mechanism can throw off ~100 watts of waste heat when the human body is in a cold room but only ~10 watts when the body is in a warm room, while the active conduction mechanism can throw off negligible heat in a cold room but up to 100-200 watts in a warm room. Thus, as the external environment warms up, the human body shifts from passive conduction to active conduction (via increased blood flow and capillary sphincter widening); if the environment becomes hotter still, or the person begins exercising, then sweating eventually comes to dominate both processes. If we now remove the active blood flow regulatory mechanism, inhibit capillary sphincter expansion, and emplace artificial tubes inside all the blood vessels and capillaries of the body (i.e., install the vasculoid), the sweating process and surface thermal radiation remain unaltered. Thus the principal changes to the human thermoregulatory system are in the passive and active conduction systems. *First,* consider the active conduction system. In a working vasculoid operating at the maximum tanker flow rate, we have ~0.133 kg (dry mass) of tankers moving around the body in ~100 sec per circuit. We assume all tankers are filled with water, for a total water mass of 0.125 kg. Diamond has a heat capacity of 519 joules/kg-K; sapphire is 728 joule/kg-K and water is 4218 joule/kg-K at 310 K. Hence, the circulating tanker fleet can transport at most ~6 watts/K of heat from the core to the periphery. Given that DT might be as small as 1 K, the "active conduction system" is essentially disabled by installation of a vasculoid that is constructed either of diamond or sapphire. *Second,* consider the passive conduction system. For a natural biological-tissue body, heat flow is $H_f = K_t / L$ = 5 watts/m^2-K. For a human body shape comprised entirely of pure diamond (K_t ~ 2000 watts/m-K at 310 K) and again taking L = 10 cm, then H_f = 20,000 watts/m^2-K. For a diamond-envasculoided human body, taking a mass of ~1.7 kg of diamond thoroughly interwoven with 68.3 kg of mostly aqueous biological tissue mass (for a standard 70 kg male body), as a crude estimate the effective heat flow becomes H_f ~ 500 watts/m^2-K, or ~100 times more thermally conductive than before. For comparison, a pure metal human form would have H_f ~ 170 watts/m^2-K for stainless steel, ~350 watts/m^2-K for lead, ~780 watts/m^2-K for iron, or ~3800 watts/m^2-K for copper. Hence a diamond-envasculoided human body would have passive conduction properties similar to those of solid metal.

This has implications for the maximum DT that can be maintained between core and periphery. Consider a human-shaped tissue-mass with half-thickness L ~ 10 cm and surface area A ~ 2 m^2, sufficiently heated from the inside to cause P ~ 100 watts (human basal rate) to flow via passive conduction from core to periphery, establishing a temperature differential DT ~ P L / A K_t ~ 10 K for natural human tissue with K_t = 0.5 watts/m-K. Upon switching to diamondoid envasculoided tissue, mean tissue thermal conductivity would rise to K_t ~ 50 watts/m-K and so DT would fall to ~0.1 K. In effect, the entire human body would become isothermal to within 100 millikelvins; even at the peak power output of 1600 watts for the human body, DT rises to just ~1.6 K. Thus, a diamond-envasculoided human body would tend to become isothermal with its surroundings very quickly (although partially offset by subcutaneous fat), a clear hazard to normal human health especially in very hot or very cold environments. The thermal equilibration time is approximately $t_{EQ} \sim L/v_{thermal} \sim 0.1$ millisec, where $v_{thermal} \sim K_t / h_{plate} C_V = 1000$ m/sec for neighboring vasculoid plates in good thermal contact with each other and having thickness $h_{plate} \sim 1$ micron, with $K_t = 2000$ watts/m-K and $C_V = 1.8 \times 10^6$ joules/m^3-K for diamond at 310 K, and taking L = 10 cm as before. This is far shorter than the typical 1-10 sec thermal response time of the purely-biological human vasculature. Substitution of sapphire for diamond significantly improves thermal performance. The thermal conductivity of synthetic sapphire may be as low as $K_t \sim 2.3$ watts/m-K for sapphire at 310 K, roughly a thousand fold lower than for diamond, when measured in the direction normal to the symmetry or optic axis (c-axis); heat capacity ($C_V = 2.9 \times 10^6$ J/m^3-K) and density (3970 kg/m^3) of sapphire are slightly higher than for diamond. Thus for a sapphire-envasculoided human body, taking a mass of ~1.9 kg of sapphire (at 25 watts/m^2-K for L = 10 cm) thoroughly interwoven with 68.1 kg of mostly aqueous biological tissue mass (at 5 watts/m^2-K), the total heat flow is just 5.5 watts/m^2-K, which differs insignificantly from natural biological tissue. At P = 100 watts, DT falls to 2 K compared to 10 K for natural tissue and 0.1 K for diamond-envasculoided tissue; $t_{EQ} \sim 1$ sec for sapphire vs. 10^{-4} sec for diamond. Two complications regarding sapphire require additional research. *First,* the thermal conductivity of sapphire may vary significantly with both composition and crystallographic orientation, a fact which might impose additional and unknown constraints on the present design. For instance, one source reports heat flow values interpolated to 310 K of 21 watts/m-K normal to the c-axis and 23 watts/m-K parallel to the c-axis; minor extrapolations of other sources to 310 K (i.e. slightly outside the exact temperature ranges measured experimentally) imply values of 2.0 and 2.3 watts/m-K for heat flows normal to the c-axis. However, all reported values for sapphire are at least two orders of magnitude more insulating than diamond. *Second,* much like diamond, the thermal conductivity of sapphire varies with temperature. For example, at ~200 K (near dry ice temperature) sapphire's thermal conductivity rises to 5 watts/m-K. At liquid nitrogen temperature (77 K), K_t soars to ~1000 watts/m-K; the peak is ~6000 watts/m-K at 35 K. (Diamond's conductivity also rises as it cools). At the other temperature extreme, sapphire's thermal conductivity rises to 3.9 watts/m-K by 523 K. Diamond thermal conductivity also varies significantly with isotopic composition (e.g., ^{12}C vs. ^{13}C); it is unknown whether similar opportunities may exist for the engineering of desired levels or patterns of thermal conductivity in isotopically-controlled sapphire. *Finally,* we note that much burn damage occurs due to a failure to dissipate heat. Selectively increasing the thermal conductivity of certain parts of the body at certain chosen times is a useful feature that could mitigate the effects of transient local heating. Note that blood is an excellent electrical conductor, and its replacement by an inorganic construct may reduce the susceptibility of the body to electrical burns under ordinary conditions. Pure single-crystal alumina (sapphire), roughly analogous to the core material of the vasculoid plates, is one of the best electrical insulators known. Pure bulk diamond is also an excellent insulator, but hydrogen-terminated diamond surface shows a high p-type surface conductivity, and slight impurities (e.g., p-type (boron) or n-type (phosphorus) or dislocations may also permit conduction, so a diamond-coated sapphire vasculoid plate should be somewhat more electrically conductive than bulk sapphire. The vasculoid is envisioned as a primarily mechanical system and so its electrical characteristics have not yet been investigated. The susceptibility of the appliance to electrostatic charge transfer, electromagnetic interference or electromagnetic pulses (EMP) is unknown.

Perhaps the single most important biocompatibility factor involving the vasculoid has been termed "vascular mechanocompatibility" by Freitas. Vasculoid nanorobots in all their forms must be as mechanically biocompatible with the vascular walls as are stents and must not produce destructive mechanical vasculopathies or disrupt the endothelial glycocalyx.

Modulation of Endothelial Phenotype and Function—The luminal surfaces of all blood and lymph vessels consist of a thin monolayer (the intima) comprised of flat, polygonal squamous endothelial cells (EC) covering a much thicker layer (the media) comprised of vascular smooth muscle cells (SMC). Under normal physiological conditions, both layers are subject (and respond) to tangential fluid shear stresses across the endothelial cell surface due to the bulk flow of blood, normal hydrostatic pressure stress acting radially on the vessel wall due to the propagation of the pressure wave, and cyclic stretch or strain due to blood vessel circumferential expansion in vivo, and thus might also be sensitive to similar mechanical stresses that may be applied by stationary or cytoambulatory intravascular nanorobots such as the vasculoid basic plates. Endothelial cells (EC) are randomly oriented in areas of low shear stress but elongated and aligned in the direction of fluid flow in regions of high shear stress. In vitro endothelial cells previously acclimatized to physiological fluid shear stresses respond to artificial changes in local fluid shear stress only very slowly, and in three stages. In the first stage, EC initially respond to the imposition of stress within 3 hours by enhancing their attachments to the substrate and to neighboring cells; the cells elongate and have more stress fibers, thicker intercellular junctions, and more apical microfilaments. In the second stage, after 6 hours the EC show constrained motility as they realign, losing their dense peripheral bands and relocating more of their microtubule organizing centers and nuclei to the upstream region of the cell. In the third stage, after 12 hours the EC become elongated cells oriented in the new apparent direction of fluid flow; stress fibers are thicker and longer, the height and thickness of intercellular junctions are higher, and the number and height of apical microfilaments are increased. This produces a new cytoskeletal organization that alters how the forces produced by fluid flow act on the cell and how the forces are transmitted to the cell interior and substrate. Physiological fluid mechanical stimuli (e.g., fluid shear stresses) are important modulators of regional endothelial phenotype and function. For example, endothelium exposed to fluid shear stress undergoes cell shape change, alignment, and microfilament network remodeling in the direction of flow (though this may be blocked via microtubule disruption using nocodazole).

Interestingly, the application of a steady laminar shear stress (a physiological stimulus) upregulates the human prostaglandin transporter (hPGT) gene at the level of transcriptional activation, whereas a comparable level of turbulent shear stress (a nonphysiological stimulus) or low stress (such as would be produced by a vascular surface coated with sessile nanorobots) does not . A few of the many quantitative experimental observations include:

1. shear stresses from 0.02-1.70 N/m^2 produce flow-induced membrane K^+ currents;
2. physiological shear stresses of 0.35-11.7 N/m^2 stimulates mitogen-activated protein kinase in a 5-min peak response time;
3. 0.04-6 N/m^2 shear stress increases inositol trisphosphate levels in human endothelial cells, with a 10-30 sec peak response time;
4. shear stresses from 0.5-1.8 N/m^2 regulate (in frequency and amplitude) oscillating K^+ currents known as spontaneous transient outward currents or STOC which are observed both in isolated bovine aortic endothelial cells and in intact endothelium; activation of STOC depends on the existence of a Ca^{++} influx and is blocked by 50 microM of Gd or is significantly reduced by 20 microM of ryanodine;
5. shear stress of 1.2 N/m^2 induces transcription factor activation over response times ranging from 0.3-2 hours;
6. Arterial shear stresses of 1.5-2.5 N/m^2 (but not a venous shear stress of 0.4 N/m^2) induce endothelial fibrinolytic protein secretion;
7. shear stress of 2 N/m^2 induces TGF-b1 transcription and production in a ~60 sec initial response time, with a sustaining increase in expression after 2 hours;
8. a shear stress of 2 N/m^2 suppresses ET-1 mRNA on confluent bovine aortic endothelial cell monolayers; these effects of shear may be completely blocked (thus allowing ET-1 to be expressed) using 875 nM of herbimycin to inhibit tyrosine kinases or 10 microM of quin 2-AM to chelate intracellular Ca^{++}, partially inhibited using 3mM of tetraethylammonium (TEA), or attenuated by elevated extracellular K^+ at 70 mM or completely inhibited by K^+ at 135 mM;
9. shear stress of 3 N/m^2 induces Ca^{++} membrane currents in a 30 sec peak response time;
10. shear stress of 6 N/m^2 applied for 12 hours causes endothelial cells to align with their longitudinal axes parallel to flow;
11. membrane hyperpolarization occurs as a function of local shear stress up

to 12.0 N/m^2, with an exponential approach to steady state in ~1 minute; the process is fully reversed once the artificial fluid flow stress is removed;

12. critical shear stress of 42 N/m^2 is the disruptive threshold for endothelial cells, inducing cell mobility; and
13. Shearing stresses of 5-100 N/m^2 occur at the contact interface when a leukocyte is adhering to or rolling on the endothelium of a venule.

Endothelial cells thus respond to sustained physiological fluid shear stresses from 0.02-100 N/m^2, spanning the range of normal arterial wall fluid shear stresses of 1.0-2.6 N/m^2 from the aorta through the capillaries and 0.14-0.63 N/m^2 for the venous circulation. By contrast, legged vasculomobile medical nanorobots may apply shear stresses during luminal anchorage or cytoambulation at velocities up to 1 cm/sec of at least 40-200 N/m^2 or higher. (Self-expanding aortic stents forcibly pulled from the vessel require an extraction force of ~400 N/m^2 assuming a 10-cm length, rising to ~1200-3600 N/m^2 for stents anchored with hooks and barbs; varying the radial force applied by stents against the vascular wall has little impact on the required extraction force.) Such shear forces, if imposed unidirectionally by large numbers of closely-packed co-ambulating nanorobots for time periods of $>10^3$ sec, may induce significant changes in shape, orientation, and physiological function in the underlying endothelial cell population. If instead these forces are applied in randomized directions varying over short time periods (<1 hour); by vasculoid nanorobots, then mechanically-induced modulation of endothelial phenotype and function should not occur. All shear forces must not be eliminated, however. A nanorobot aggregate that shields vascular cells from fluid shear for an extended time may induce those cells to revert to their flow-unstressed phenotype or to undergo apoptosis. Endothelial cells cultured in the absence of shear stress tend to become dedifferentiated. In one study, after blood shear was artificially reduced near a wound lesion for 24 hours the local endothelial cells became less elongated, contained fewer central microfilament bundles, and exhibited a slower repair process. In another study, vein grafts removed from the higher-shear arterial circulation and reimplanted in the lower-shear venous circulation of the same animal showed regression of intimal hyperplasia and medial rethickening in 14 days, apparently due to induction of smooth muscle cell apoptosis by a reduction in pressure or flow forces. Endothelial cells can also respond to persistent static overstretching in many ways, up to and including apoptosis. For instance, hypertension caused by hydrostatic edema can induce apoptosis in capillary EC. Additionally, vascular wall cells respond to lateral stretch forces due to cyclical blood vessel expansion in vivo. For example, in one experiment bovine aortic endothelial cells were seeded to confluence on a flexible membrane to which cyclic strain was then applied at 1 Hz (0.5 sec strain, 0.5 sec relaxation) for 0-60 min. After 15 minutes of this cyclic stretching, there was an increase in adenylyl cyclase (AC), cAMP, and protein kinase A (PKA) activity of 1.5-2.2 times at 10 per cent average strain as compared to unstretched cells, but there was no activity increase at 6 per cent strain—evidently, cyclic strain activates the AC signal transduction pathway in endothelial cells by exceeding a strain threshold, thus stimulating the expression of genes containing cAMP-responsive promoter elements. Stretch-activated cation channels in bovine aortic EC are inhibited by $GdCl_3$ at 10 microM. Human umbilical EC subjected to a 3-sec stretch pulse show an intracellular rapid-increase Ca^{++} spike, followed by a (ryanodine-inhibitable) slow-decline, due to Ca^{++} entry into the cell through stretch-activated channels; Mn^{++} also permeates mechanosensitive channels (but not Ca^{++} channels) and enters the intracellular space immediately after an application of mechanical stretch. Cyclic strains of 10 per cent at 1 Hz induce intracellular increases in Ca^{++}, diacylglycerol, inositol trisphosphate and protein kinase C (PKC) in peak response times of 10-35 sec, often sustained for up to ~500 sec, and induce transcription factor activation over response times ranging from 0.25-24 hours. Several endothelial cytokines are induced by cyclic mechanical stretch and cyclic mechanical strain modulates tissue factor activity differently in endothelial cells originating from different tissues.

Similarly, bovine aortic smooth muscle cells (SMCs) seeded on a silastic membrane and subjected to cyclic strains up to 24 per cent enhanced SMC proliferation at any strain level, although SMC under high strain (7-24 per cent) showed more proliferation than SMC at low strain (0-7 per cent) in this experiment. High-strained SMC aligned themselves perpendicular to the strain gradient, whereas low-strained SMC remained aligned randomly; PKA activity and CRE (cAMP response element) binding protein levels increased for highly strained cells, compared to low-strained cells. Other experiments have found that small mechanical strains of 1-4 per cent at 1 Hz applied to human vascular smooth muscle cells can inhibit intracellular PDGF- or TNFa-induced synthesis of matrix metalloproteinase (MMP)-1; that saphenous vein SMC distention by 0.5 atm pressure subsequently elevates cell apoptosis; that cyclic mechanical strain at normal physiological levels decreases the DNA synthesis of vascular smooth muscle cells, holding SMC proliferation to a low level; that 1 Hz, 10 per cent cyclic strain on SMC activate tyrosine phosphorylation and PKC, PKA, and cAMP pathways over response times from 10 sec to 30 min and that vascular SMC exhibited stretch-induced apoptosis when subjected to cyclic 20 per cent elongation stretching at 0.5 Hz for 6 hours. Consequently, medical nanorobot aggregates which shield the vasculature from normal cyclical strains might elicit excess growth of vascular smooth muscle cells, which growth is normally held in check by the rhythmic stretching from the arterial pulse. Intravascular nanorobot aggregates that apply cyclic mechanical strains exceeding a few percent might encourage increased SMC proliferation and activate mechanosensitive and stretch-activated channels in EC, along with cellular realignment and subsequent SMC apoptosis at the highest strain levels. In 2002 it was unknown whether high frequency (>KHz) cyclic mechanical strains likely to be employed by vasculomobile medical nanorobots would have biological effects similar to or different from those described above for low-frequency cyclic strains, excepting certain specialized mechanoreceptor cells such as the cochlear stereocilia, other hair cells and somatosensory neurons, since most mechanical cell stimulation experiments have been conducted at low frequencies. Unrecognized effects that might be triggered by high-frequency cyclic strains cannot be ruled out. However, given the relative safety of procedures involving intravascular ultrasound with its low complication rate (e.g., only 1.1 per cent, including spasms, vessel dissection and guidewire entrapment using frequencies as high as 10-20 MHz, it seems improbable that KHz or MHz acoustic waves of the intensities that might be employed by medical nanorobots for communication or power supply will damage the luminal vascular walls. Continuous low-power ul trasound exposures exceeding ~10^4 sec are considered safe but there are few studies on the safety of long-duration chronic exposures. (Interestingly, relatively high-intensity intravascular ultrasound has been used to dissolve occlusive platelet-rich thrombi safely and effectively in myocardial infarctions and in restenosed stents.) If necessary, large wave motions and pressure patterns that are characteristic of normal blood flow can in principle be simulated by the vasculoid plate array by a combination of motile ciliary activity and periodic manipulations of interplate bumpers. It is likely that the vasculoid appliance will need to control smooth muscle cell proliferation in the simplest case releasing specific cytokines into the vasculoid-endothelial space. Such factors may include known SMC proliferation promoters such as thrombin (esp. alpha-thrombin), PDGFs (esp. PDGF-AA), FGF (esp. basic FGF), HBEGF (heparin binding epidermal growth factor), TGFb (transforming growth factor-beta) at low concentrations, angiotensin II, thrombospondin-1 (stretch/ tension), and known SMC proliferation inhibitors such as heparin/heparan sulfate, TGFb (transforming growth factor-beta) at high concentrations, nitric oxide, prostaglandins, calcium antagonists, agonists that activate guanylate and adenylate cyclases, inhibitors of angiotensin-converting enzyme, interferon gamma, 18-beta-estradiol, sodium salicylate, and the topoisomerase I inhibitor topotecan. Adult arterial walls contain both differentiated and immature SMCs. R. Bradbury notes that further research is needed regarding how SMCs handle conflict resolution between "divide" and "don't divide" signals that it may be receiving from both internal and external sensors. Given the large number of signals that SMCs currently respond to, it seems highly likely that the vasculoid can "manage" them, along with other secretion products of SMCs that play important roles in the prevention of vascular disease, such as extracellular superoxide dismutase.

Vascular Response to Stenting—Mechanical biocompatibility must also be demonstrated by intravascular nanorobots that are intended to remain in long-term near-contact with blood vessel walls. A related medical analog is the vascular stent—a flexible metal coil or open-mesh tube that is surgically inserted into a narrowed artery, expanded, and pressed into the vascular wall at up to 10-20 atm pressure, in order to ensure long-term local vascular patency by providing a scaffold to hold the artery open. Within 4 days, SMC begin to appear in the intima. After a few months the stent is completely overgrown with new endothelium, forming a neointima, although the media is usually compressed with smooth muscle cell atrophy in all stented regions; stenosis is prevented in vessels 10 mm or greater in diameter but is not precluded in vessels smaller than 6 mm. Histologically, in-stent restenosis appears to derive almost exclusively from neointimal hyperplasia, which appears more abundant following stent implantation than balloon angioplasty and in stents of greater stent length and smaller vessel caliber (or inadequate stent expansion). Restenosis occurs in 22-46 per cent of all stents emplaced within 6-12 months), in some cases requiring the insertion of a second stent into the first, and varies according to the material used. In one experiment, the thickness of the neointimal layer formed over wire-mesh stents placed in canine aortas was 83.9 microns thick for gold, 103.6 microns for stainless steel, 115 microns for Teflon, 209.6 microns for silicone, and 228.6 microns for silver; a copper stent produced severe erosion of the vessel wall, marked thrombus formation, and aortic rupture. Stent surface coating and texture also affects leukocyte-platelet aggregation and platelet activation. Improved prospects are reported for diamondoid stents. For instance, diamond-coated stents produced by Phytis Corp are inserted at 16 atm pressure (vs. 2-3 atm normally for stents) and yet do not dislodge surface (thrombogenic) antigens and selectins: "The results show that in none of the control systems (systems without stents) [could] a change in glycoprotein expression (i.e. all antigens) ... be detected. With the exception of one measurement also the structural epitopes (CD 41a, CD 42b) show no significant changes during the test period. The reason is most probably that the shearing strength of the system was too weak or the expression of the antigens was too strong. Whereas for the thrombocyte activation, remarkable difference in the expression of CD 62p and CD 63 could be detected. The thrombocyte activity marker CD 62p is reduced in diamond-like coated stents and diamond-like coated stents with heparin compared with uncoated stents." However, stent devices are far from ideally mechanocompatible with blood vessel walls. For example, stents placed endovascularly in dog aorta for 4-45 weeks and then examined histologically show medial atrophy, intimal hyperplasia (tissue ingrowth), and proliferation of the vasa vasorum (the microvasculature of the aorta) more prominently for covered stents than for bare stents, probably due to hypoxia in the aortic wall. Cellular proliferation is highest when the artery wall is most hypoxic; but vasculoid nanorobot aggregates (e.g., basic plates) that entirely cover the vascular endothelium can precisely regulate oxygenation of the underlying tissue (using data from oxygen sensors in the plate wall contacting the endothelium), thus largely eliminating the possibility of hypoxia. Stentlike vascular-coating structures also may be able to inhibit stenosis due to vascular smooth muscle proliferation, migration, and neointima formation, without inducing apoptosis, by releasing the topoisomerase I SMC-proliferation inhibitor topotecan in a localized 20 min exposure, or by using other similar drugs.

Nanorobotic Destructive Vasculopathies—The physical configurations or activities of medical nanorobotic aggregates could in some circumstances be destructive of vascular tissue. Owens and Clowes point out that the severity of arterial injury is important in determining the ultimate pathophysiologic response and describe a classification system based on the immediate histologic effect of the injury:

Type I injuries involve no significant loss of the vessel's basic cellular architecture, although there may be a slight change in endothelial architecture and associated cellular adhesion. Examples include the fatty streak (an early atherosclerotic lesion), hemodynamic factors and flow disturbances which produce, at most, only a modification of the established cellular architecture.

Type II injuries involve loss of the endothelial layer, perhaps inducing platelets to adhere and begin forming a thrombus at the area of loss, but the internal elastic lamina remains intact and there is little or no damage to the media. Examples include injuries incurred during simple arterial catheterization, endovascular procedures, vein graft preparation, or gentle filament-induced endothelial denudation of the carotid artery in a rat model.

Type III injuries involve transmural damage in which the endothelium is removed, the internal elastic lamina is often disrupted, and a significant portion of the medial cells are killed; platelets deposit and a thrombus forms at the site of endothelial loss, and an inflammatory response (vasculitis) including intimal hyperplasia is initiated within the vessel wall. Examples include spontaneous vascular dissection and various forms of surgical repair or reconstruction such as balloon angioplasty, endarterectomy and atherectomy.

Medical nanorobot device and mission designs should always seek to avoid *Type II* and *Type III* injuries, although in some special circumstances the potential even for *Type III* injuries may be inescapable. Destructive mechanical vasculopathies that might be caused by vasculoid nanorobots may be classified as ulcerative, lacerative, or concussive.

Nanorobotic Ulcerative Vasculopathy—Pressure ulcers are normally caused by a prolonged mechanical pressure against epidermal tissues (e.g., the skin of a person who is lying down; decubitus ulcer, or bedsores), typically at sites over bony or cartilaginous prominences including sacrum, hips, elbows, heels and ankles. The combination of pressure, shearing forces, friction and moisture leads to tissue death due to a lack of adequate blood supply; if untreated, the ulcer progresses from a simple erosion to complete involvement of the dermal deep layers, eventually spreading to the underlying muscle and bone tissue. In rare cases, mechanical frictional stimulation of the skin can precipitate systemic cutaneous necrosis and calciphylaxis, a state of induced tissue sensitivity characterized by calcification of tissues. Stercoral ulcers are caused by the necrosis of intestinal epithelium due to the pressure of impacted feces. Fluid mattresses can greatly reduce pressure ulcers in long-duration surgeries and pressure-relieving surfaces have been investigated for surgical patients, wheelchair users, and for other circumstances) It is generally recommended that the interior surface should employ materials having roughly the same mechanical properties as the enclosed tissue, and the applied interfacial pressures should be reduced to below 1 psi or ~ 50 mmHg. Similarly, a macroscale nanorobot aggregate such as a vasculoid appliance may cause luminal vascular ulceration by prolonged mechanical pressure against intimal tissues, similar to epithelial pressure ulcers, necrotizing vasculitis, or pressure necrosis. For example, mechanical stretch induces apoptosis in mammalian cardiomyocytes and hypertension caused by hydrostatic edema can induce apoptosis in capillary endothelial cells. Another example is IUD-induced metrorrhagia (nonmenstrual uterine bleeding), wherein the IUD (intra-uterine device) elicits a vascular reaction most pronounced in the endometrium adjacent to the device and includes increased vascularity, degeneration with defect formation, congestion, and poor hemostatic responsiveness to increased vascular permeability and damage, leading to interstitial hemorrhage due to vascular damage from mechanical stress transmitted by the IUD through the endometrium to its vascular network. Indwelling catheters can rest very snugly against the vascular walls without complication for brief periods. A biological-like interface may reduce ulceration in longer-term missions. In one study, a stented aortic graft was placed endovascularly in the native aorta of male sheep, and a histological examination 6 months later found good incorporation of the graft with no pressure necrosis, although there was a foreign body reaction around the graft and an organized blood clot was noted between the graft and the aortic wall (the inner, but not the outer, aortal surfaces would be expected to have clotting-suppressive properties). However, long-duration nanoaggregates (such as vasculoid plates) that must maintain close contact with endothelium should employ a mechanically compliant coating having properties similar to extracellular matrix. All such linkages should be not just immunocompatible but also mechanocompatible, possessing equivalent elasticity or mechanical compliance as the underlying tissue to which attachment must be secured. With conventional implants, compliance design may include assessments of circumferential compliance (measurement of changes in vessel diameter over a complete cardiac cycle, including pressure-radius curves, dynamic compliance, and mechanical hysteresis effects), longitudinal compliance (elasticity of selected lengths of the vascular system, including any localized stiffening), tubular compliance (imparity of elasticity between a prosthetic conduit and the native artery, elastic energy reservoiring, and pulsatile energy losses due to interfacial impedance mismatches), and anastomotic compliance (suture line anastomotic compliance mismatch and the para-anastomotic hypercompliance zone, localized regions of excessive mechanical stress, and cyclic stretch effects on replication of vascular SMC and extracellular matrix). A mismatch in mechanical properties between relatively compliant arteries and less-compliant metallic stents and tissue grafts has been thought to influence patency and pseudointimal hyperplasia. Larger more central arteries are more compliant than the distal small caliber arteries, wall shear stress from blood flow differs on either side of a curving vessel and the stress is out of phase with the pulsing circumferential stretch strain, and compliance mismatch between host artery and prosthetic graft may promote subintimal hyperplasia. Analogous compliance issues may be assessed once the static and dynamic stress patterns in the vasculoid transport system are more precisely known. Post-installation vascular conditioning can be maintained on a permanent basis because the vasculoid appliance exercises precise control over the transport of most metabolic and cellular traffic—e.g., of cholesterol, leukocytes and platelets, the principal participants in the arteriosclerogenic process, or of various circulating pluripotent stem cells and their activating factors.

Nanorobotic Lacerative Vasculopathy—Individual vasculomobile nanorobots or nanorobotic aggregates may occasionally scratch, scrape, or gouge the vascular luminal surface, causing partial or complete loss of local endothelium (*Type II* damage), a form of mechanical vasculitis or capillaritis, particularly during installation of the vasculoid appliance. Since the typical dimensions of nanorobotic vasculoid components approximate the endothelial thickness, transmural *Type III* damage to the media is unlikely. Turnover studies of rat endothelium show that (a) injured endothelium can recover an area one cell wide (~1000 micron^2) in ~3 hours, (b) the natural loss rate is ~0.1 per cent of endothelial cell area per day (~1 micron^2/day), and (c) the steady-state vascular denudation area is ~0.125 micron^2/cell.

Smooth nanorobot hulls, boundary layer effects and low fluid flow velocity throughout most of the vasculature during installation should ensure that major "sandblasting" type erosion is unlikely to occur inside human blood vessels even at the highest nanocrits consistent with continuous flow. Free-floating nanorobots that collide with blood vessel walls (given the no-slip condition at the wall) produce minimal shear forces, on the order of <~0.1 N/m^2—this is less than the 1.0-2.6 N/m^2 shear forces normally encountered in arteries and capillaries due to normal blood flow and the 0.14-0.63 N/m^2 shear forces in veins, but may be sufficient to cause a small biological response from the vascular endothelium. Applying the maximum possible bloodstream velocity of 1 m/sec to the impact-scratch relation, it is clear that particle-wall collisions should produce only harmless submicron nicks even in the most turbulent arteries. Nevertheless, some caution is warranted because natural endothelial cell wounding of ~1-18 per cent of all cells, possibly erosionally-derived, has been observed in rat aorta. Erosion of cultured fibroblast monolayers (simulating the vascular endothelium) using MHz ultrasound at acoustic pressures of ~10^6 N/m^2 is enhanced by the presence of a microbubble (particulate) contrast agent. Injection of crystalloid cardioplegic solutions into canine hearts at pressures >110 mmHg and at peak flow rates >25 ml/sec also causes a higher incidence of mechanical-physical trauma to the vascular endothelium and the endocardium. In another unusual case, intravenous self-injection by a drug abuser of dissolved tablets containing microcrystalline cellulose as filler material produced numerous microcrystalline cellulose pulmonary emboli, intravascular foreign body granulomas, focal necrosis and edema of the pulmonary parenchyma, and fatal vascular destruction. Endothelial abrasion alone may not stimulate neointimal thickening but inevitably must involve some endothelial cell loss and other biological responses. For example, mechanically scraping cultured endothelial cells causes growth factor to be released within 5 minutes, not abating for at least 24 hours thereafter, due to plasma membrane disruption. In the case of vascular dissection, a piece of the endothelium peels up, making an intimal flap that defines regions of true and false lumina, and sometimes may induce an intramural hematoma in the aortic wall. Endothelial cells mechanically damaged with a razor blade activate extracellular-signal-regulated kinases within ~300 sec, releasing fibroblast growth factor (FGF-2) which in turn induces intimal hyperplasia. Nanorobots which detect FGF-2 are alerted that mechanical endothelial injury has taken place; by absorbing the cytokine using molecular sorting rotors, the hyperplasia signal may be suppressed by a nanorobot, if desired. However, shear-induced endothelial denudation of healthy canine arterial endothelium appears not to occur at shear stresses up to at least 200 N/m^2. The role of erythrocyte collisions with vascular walls on the detachment rate of endothelial cells is just starting to be seriously investigated. K. Clements suggests that some provision might be also be made for emergency plate disconnects in certain unusual accident situations, e.g., where the user has irreversibly caught his hand in a machine applying superior force, and now risks not just amputation of the original biological limb, but also faces either (1) forcible extraction of the diamondoid appliance from the remaining biological tissues, or (2) progressive amputation of additional biological tissues because the vasculoid network has become caught in the machine.

Nanorobotic Concussive Vasculopathy—If a patient experiences significant external crushing or concussive forces, resident medical nanorobots that are present in small numbers can simply move out of the way, as described previously by Freitas in connection with the risks of dental grinding. In the case of macroscale intravascular nanoaggregates, there are several additional risks that should be avoided in specific appliance designs. *First,* there is the possibility that a sudden mechanical external tissue compression could push macroscale nanorobotic aggregates through the soft tissues, causing deep tissue penetrations, perforations, or other serious mechanical trauma. Similarly, because the vasculoid materials may have higher density than the surrounding biological tissues, very high accelerations could produce effects on those tissues that would be not unlike pushing gelatin through a metal wire strainer. Simple activities such as hand clapping or fistfights are unlikely to produce the high accelerations required for serious damage, but this risk should be quantified in future studies. Possibly relevant analogies in the medical literature include:

(1) Ulnar artery erosion, thromboemboli, digital ischemia and skin necrosis from a glass foreign body in a patient's hand.

(2) Tantalum coil stent damage that was induced or aggravated by intravascular ultrasound inside a coronary artery.

(3) A chronic indwelling catheter that led to erosion and rupture of the anterior wall of the right ventricle, producing a near-exsanguinating hemorrhage.

(4) Cardiac perforation by a subclavian catheter.

(5) Pulmonary artery catheter-induced right ventricular perforation during coronary artery bypass surgery.

(6) An ICD patch that migrated and perforated the right ventricular cavity.

(7) A stent that migrated to an oblique position across the aorta, producing a 7-cm pseudoaneurysm after 3 years.

(8) Catheter-induced pulmonary artery rupture (a well-recognized complication of invasive monitoring) that often leads to fatal hemorrhage.

(9) Femoral artery catheterization trauma producing hematoma, pseudoaneurysms and arteriovenous fistulas of the femoral vessel.

(10) Iatrogenic subclavian artery injury due to central venous catheterization.

(11) Repeated and prolonged vein catheterization that led to subsequent stenosis (presumably due to luminal vascular mechanical damage).

(12) High-pressure injection injury that induced inflammation and foreign body granulomatous reaction, progressing to necrosis.

(13) Mechanical tearing of arteries due to overstretching.

(14) Spontaneous coronary artery dissection (mechanical arterial wall failure).

(Most of these cases pertain to injury from objects much stiffer and larger than vasculoid components, or are due instead to intrinsic vessel dysfunction (as in spontaneous arterial dissection.)

Second, a sudden external tissue compression could force nanorobotic aggregates into physical contact with neighboring nanoaggregates, possibly causing major structural damage or fragmentation of the devices. This risk increases as the nanodevices become more densely packed, especially along the crushing axis. Nanoorgans (as well as looser aggregates) can be crushed if sufficient force or mechanical shock is applied. Again, a few possibly relevant analogies from the medical literature include:

(1) External compression of emplaced stents that produced premature stenosis.

(2) A transabdominal teflon stent that broke intraperitoneally during tuboplasty procedure.

(3) A strongly-beating heart that sheared off a pericardial drainage catheter.

(4) A Hickman catheter that suffered rupture and embolization during normal use.

(5) An indwelling catheter that fractured and a distal remnant embolized to the right ventricular outflow tract and main pulmonary artery, precipitating cardiopulmonary near-collapse.

(6) A catheter embolism that was produced when a catheterized patient engaged in power training exercises, externally crushing the catheter, although no symptoms or complications accompanied this event.

(7) Spontaneous fracture of indwelling venous catheter, leading to vascular leakage.

(8) Other instances of catheter fracture and embolism, including one case that led to cardiac arrest.

(Here, too, many of these cases might differ substantially from the vasculoid, since the catheters were made from a bulk material that depended on its integrity for function while the vasculoid would consist of a large number of semi-independent components capable of restoring functional connections on their own.)

Third, there is a small risk that poor device design, poor mission design, or a loss of control might cause nanoaggregates to operate in a dangerous manner, causing macroscale concussive injury to biological tissues. For example, iatrogenic vascular trauma caused by intra-aortic balloon pumps is well-known. In one case, balloon expansion during angiography ruptured the pulmonary artery. In a canine model, an aortic valve balloon dilation to 5-12 atm pressure produced valve leaflet connective-tissue injury and hemorrhage. In an experiment with canine intestine, excessive lymph pressures (>850-1630 N/m^2) produced artificially in the central lacteals caused fluid to leak out of the villi and caused intestinal epithelial cells to be shed into the intestinal lumen. These risks can be largely avoided or at least minimized by good design and experience with related systems, initially in animal models.

Mechanical Interactions with Glycocalyx—Endothelial cells are surrounded by a well-developed extracellular glycocalyx. If this outer margin is traumatized, receptor sites and fibronectin may be exposed which could then become available for bacterial adhesion. Nanorobots that rely upon absorption of local oxygen and glucose for their power supply or whose missions include extensive small-molecule exchanges with the environment may have ~10^4-10^5 molecular sorting rotors embedded in their exterior surfaces. These spinning sorting rotors are unlikely to cause direct physical damage to cell surfaces for several reasons. First, rotors are atomically smooth and recessed into the housing, reducing physical contact with colliding surfaces and eliminating potential nucleation sites that may trigger thrombogenesis, gas embolus formation, or foaming. Second, only a small fraction of all available sorting rotors may be actively spinning at one time, further reducing the likelihood of physical trauma. And such limited contact, when it occurs, should be relatively benign: Maximum rotor rim velocity of 2.6 mm/sec is less than 1 per cent of mean aortic blood velocity and lies only slightly above maximum capillary flow speed.

Many disease processes are known that involve damage to the glycocalyx, including some bacteria that phagocytose or otherwise destroy the cellular glycocalyx during an infection. Damage to the glycocalyx creates conditions that favor the binding of immune complexes, complement activation, and intravascular coagulation, with loss of gradients between blood and parenchyma; desialylated glycocalyx of endothelium also allows an increased rate of endothelial cell detachment from arterial walls. Could the glycocalyx strands present at all tissue and nontissue cells surfaces get trimmed, even by a recessed sorting rotor? Nanorobot sorting rotor binding sites for small molecules (<20 atoms) involve pockets measuring <2.7 nm in diameter, too small to physically accommodate the 10-20 nm thick plasma membrane or the main body of the glycocalyx projections typically measuring 5-8 nm thick and 100-200 nm long, consisting of glycoproteins comprised of 10,000 atoms or more. While an occasional sugar residue may get clipped, binding sites can be designed for maximum steric incompatibility with glycocalyx glycoproteins and proteoglycans, further minimizing the opportunities for trimming. Note that clipping a covalent C-C, C-O, or C-N bond probably requires a clipping energy >500 zJ/molecule, but sorting rotors designed to pump against pressures of ~30,000 atm can only apply ~100 zJ/molecule (i.e., per binding site) so an accidentally-bound glycocalyx moiety seems more likely to jam the rotor than to be clipped off by the rotor. If this happens, the result may be a glycocalyx-tethered nanorobot, in which case a rotor-dejamming protocol will be required to free the trapped nanorobot. Natural rates of glycocalyx damage are just starting to be quantified, and many tissue cells replace their glycocalyx or are retired in times ranging from 10^3-10^6 sec. For example: Schistosome parasites can shed some tegument-bound complexes in only ~1200 sec to 3600 sec; plasma membrane turnover rate is ~1800 sec for macrophage and ~5400 sec for fibroblast; cholesterol turnover rate in RBC membrane is ~7200 sec membrane phospholipid half-life averages ~10,000 sec; neutrophil lifespan in blood is ~11,000 sec; enterocyte glycocalyx is renewed in 14,000-22,000 sec, as vesicles with adhered bacteria are expelled into the lumen of small and large intestine; some schistosome membrane antigen turnover may require from 68,000 to 160,000-430,000 sec; typical protein turnover half-life is ~200,000 sec; cell turnover time is ~86,000 sec in gastric body, ~200,000 sec for duodenal epithelium, ~240,000 sec for ileal epithelium, and ~400,000 sec for gastric fundus; neutrophil lifespan in tissue is ~260,000 sec; glycocalyx turnover in rat uterine epithelial cells is ~430,000 sec; and platelet lifespan is ~860,000 sec). As the cell coat is a secretion product incorporated into the plasma membrane that undergoes continuous renewal, any trimmed glycocalyx glycoproteins from tissue cells would be rapidly replaced via biosynthesis in the ribosomes of the endoplasmic reticulum, followed by final assembly with the oligosaccharide moiety in the Golgi complex and subsequent export to the plasma membrane. Glycoprotein strands or stray sugar residues released into the extracellular medium as a consequence of such trimming are nonimmunogenic and would be quickly metabolized, although it is possible that nearby parasites could absorb this released material onto their surface, affording themselves some camouflage protection against natural host immune defenses but little protection against vasculoid defenses since parasite antigens should still be visible at cell surfaces.

A nanorobotic aggregate covering a macroscale area of the capillary luminal surface may reduce the normal flow of plasma water and other substances that leaves the circulation via ultrafiltration, unless the necessary water is replaced by the nanosystem.

The plasma water flow helps to remove waste products from the extracellular spaces around tissue cells, a function that could be compromised by the shielding presence of the nanoaggregate unless the aggregate replaces this flow with water transported through or around the device, by various means. Consequently, both lymph volume and the gross water flows between tissues may be affected by vasculoidization, given that the vasculoid substitutes encapsulated fluid transport in place of bulk-flow and diffusive fluid transport in the principal global materials distribution system of the human body. More specifically, ~20 liters/day (0.23 cm^3/sec) of plasma water exit the natural circulation via ultrafiltration through leaky capillaries, of which 18 liters/day are reabsorbed after passing through the lymphatic capillaries and back into the venous loops, leaving a net flow of ~2 liters/day to pass onward through the lymphatic system. The maximum vasculoid water transport rate has been scaled to 0.60 cm^3/sec or ~52 liters/day, which is the most that the water tanker subsystem can handle. Under normal conditions only ~4 per cent of the water tanker fleet is in use, which still allows the transit of at least 0.024 cm^3/sec or ~2.1 liters/day of water which if not re-imported by the vasculoid on the venous side of the capillary bed is forced to enter the lymphatic system, maintaining normal flow and volume. This flow should be sufficient because the direct removal from the intercellular fluids of pure waterborne electrolytes, solutes, and other physiological substances by the vasculoid (through the plates) can create pericellular concentration gradients sufficient to ensure injection and removal of such substances from the vicinity of the bathed cells. This should greatly reduce the gross intercellular fluid flow rate required to maintain proper physiological conditions. One minor additional concern might arise because lymph is moved primarily by peristaltic one-way valving. Envasculoided tissues adjacent to lymphatic capillaries may have slightly higher mechanical stiffness, hence may transmit somewhat less peristaltic action resulting in reduced lymphatic flow rates. But lymph fluids transport solutes at far below maximum concentration—for example, lymph normally contains ~0.003 gm/cm^3 NaCl, two orders of magnitude less than the ~0.36 gm/cm^3 maximum solubility at 310 K– and bacterial entry will be largely precluded by the appliance, so small reductions in lymph flow rates are probably tolerable. Lymphatic venules do exhibit small-amplitude pulsations on their own, and it may be desirable for other reasons to simulate the stiffness and motion (including motion resulting from blood flow) of pre-vasculoid tissue. Lymph movement also may be replaced by an optional "lymphovasculoid" appliance. Similar considerations and conclusions apply to other fluid spaces within the body that communicate, directly or indirectly, with the blood, including the choroid plexus and cerebrospinal fluid, as well as the pericardial, pleural, peritoneal, synovial, and intraocular fluid spaces.

CVD (chemical vapor deposition) diamond coatings are said to have "low immunoreactivity" and there were no reports of diamond immunogenicity in the medical literature as of 2001. Indeed, diamond is often used as an experimental control because it is so chemically inert and biologically inactive. Pure sapphire also appears fairly nonimmunogenic, although similar hydrophilic surfaces do adsorb immunoglobulin IgG; single-crystal sapphire has excellent biological inertness and chemical stability. However, both diamond nanoparticles and various soluble aluminum salts (e.g., alum, aluminum hydroxide, and aluminum phosphate) have been shown to serve as adjuvants which enhance vaccines or immune system responses to foreign antigens. Exposed to water, the polished single-crystal a-alumina (0001) surface elicits a hydration reaction, with a water vapor pressure of ~1 torr sufficient to fully hydroxylate the surface. Alumina is corrosion-resistant because it exists in the highest oxidation state that aluminum metal can possess, and has the potential for microstructural control of the interface (with tissue) without formation of toxic corrosion products. Yet it is also known that a-alumina is very slightly soluble in highly acidic or alkaline aqueous environments. Since Al^{+++} ions can produce "dialysis dementia" and are generally considered toxic, it is of interest to determine whether or not these ions can leach into the body from alumina implants or sapphire nanorobots. Early studies in the 1970s found no movement of known contaminants into the surrounding tissue from sintered alumina implants inserted into the iliac crests (hip bones) and mandibles of rabbits. During the 1980s and 1990s, small increases in blood aluminum concentrations were demonstrated in smelter workers, though the potential exposure level is several orders of magnitude greater for body uptake of more soluble aluminum compounds used as food additives, as antacid medication, or from food packaging materials and cooking utensils. In 1990, Lewandowska-Szumiel and Komender investigated aluminum release from an alumina bioceramic during standardized biocompatibility testing in an animal experiment. Alumina implants introduced into rat femurs and guinea-pig mandibles and then removed 6-8 months later were found to be well tolerated, and no changes in the surfaces of the removed implants were observed under SEM examination. The researchers decided to compare the aluminum content of the femurs of experimental and control rats using atomic absorption spectroscopy, and discovered that the level of aluminum was higher in the bones of the experimental animals. In 1991, Arvidson et al investigated the corrosion resistance of single-crystal sapphire implants with respect to the release of aluminum ions, and found no ions in the test solutions. The next year, Christel reported that alumina exhibited greater bioinertness than all other implant materials currently available for joint replacement, and that no lymphocyte or plasma-cell infiltration into joint implants is observed "because of the absence of soluble component release." However, two studies in the early 1990s found some detectable aluminum ion release, so more research is clearly required on this issue. In any case, a thin veneer of diamond on sapphire should suffice to prevent aluminum ion release in vivo. An allergic reaction or "hypersensitivity" is an acquired and abnormal immune system response to a substance, called an allergen, that normally does not cause a reaction. An allergy requires an initial exposure to an allergen which produces sensitization to it subsequent contact with the allergen then results in a broad range of inflammatory responses. Sapphire or alumina ceramic is considered nonallergenic—ceramic coatings are used to eliminate metal allergies on implant surfaces and hypersensitivity to oral ceramic is reported only rarely. There are no reports of allergenicity for diamond, sapphire, fullerenes, or other probable diamondoid nanorobot exterior materials and such allergenicity appears unlikely, but experiments should be done to positively confirm this expectation. The immune system could also react to small subcomponents like cilia, especially if they form aggregates with proteins, possibly requiring clonal deletion or tolerization to deal with such issues.

Could the vasculoid surface in contact with the vascular endothelium trigger general inflammation in the human body? One early experiment to determine the inflammatory effects of various implant substances placed subdermally into rat paws found that an injection of 2-10 mg/cm^3 (10-20 micron particles at 10^5-10^6 particles/cm^3) of natural diamond powder suspension caused a slight increase in volume of the treated paw relative to the control paw. However, the edematous effect subsided after 30-60 minutes at both concentrations of injected diamond powder employed. Another experiment at the same laboratory found that intraarticulate injection of diamond powder was not phlogistic (i.e., no erythematous or edematous changes) in rabbit bone joints and produced no inflammation. Diamond particles are traditionally regarded as biologically inert and noninflammatory for neutrophils and are typically used as experimental null controls. CVD diamond and DLC diamond surfaces elicit minimal or no inflammatory response, and atomically smooth diamond may perform even better. Diamond particles are said to have little or no surface charge but unmodified graphene surfaces readily acquire negative charges in aqueous suspension, so experiments are needed to determine if negatively charged fullerenes or other diamondoid substances can contact-activate Hageman factor or kallikrein and trigger an inflammation reaction. Experiments with sapphire have generally found no serious inflammation in soft tissues or bony tissues, or only mild reactions, though there are a few modest exceptions including a brief acute inflammatory response.

Blood coagulation involves a complex series of reactions in which various proteins are enzymatically activated in a sequential manner, transforming liquid blood into a gel-like clot which is then stabilized to form a thrombus (clot) consisting of platelets, fibrin, and red cells. The series of reactions is classically divided into two pathways—extrinsic and intrinsic —involving more than a dozen factors that converge on a single common final pathway, resulting in clot formation. Since, these factors are carried by the blood, the vasculoid can regulate their local concentration and thus, prevent thrombogenic pathways from proceeding to completion. Additionally, platelets must undergo adhesion and activation for coagulation to occur. The adhesion of platelets to exposed collagen in injured blood vessels is mediated by a bridging molecule called von Willebrand's factor that is secreted by endothelial cells into plasma, which prevents platelets from detaching under the high shearing stresses developed near vessel walls. The activation of normally quiescent platelets is a complex phenomenon that includes changes in cell shape, increased movement, release of the contents of their granules (containing nucleotidyl phosphates, serotonin, various factors, enzymes and plasma proteins), and aggregation. The most potent activator of platelets in vivo is thrombin, which interacts with a receptor on the platelet plasma membrane, followed by transmembrane signaling and subsequent activation of the cell. Collagen is the other most important platelet activator; ADP can stimulate aggregation but not granule release. In principle, the blood-contacting surfaces of a nanoorgan, or of nanorobots in sufficient bloodstream numbers and concentrations, could activate platelets (and thus either of the two coagulation pathways), but careful choices of materials and of allowable mechanical motions should reduce or eliminate inherent nanodevice thrombogenicity. For example, DLC diamond-coated stents, heart valves and other blood-contacting LVAD surfaces or substrates generally show reduced thrombogenicity and no platelet activation. Sapphire (alumina ceramic) has low thrombogenicity and both platelet adhesion and activation are low. Hemolysis is near-zero for diamond and alumina powders. Additionally, in the vasculoid plateletogenesis and release of platelets could be actively regulated and reduced to minimal levels. Future experiments must determine if ordinary diamondoid surfaces will have to be supplemented with additional antithrombogenic coatings in order to achieve vasculoid performance objectives. If such coatings are required, one simple possibility is surface-immobilized heparin, a ~15 kD straight-chain anionic (acidic) mucopolysaccharide (glycosaminoglycan) that forms polymers of various lengths. Heparin, first discovered in 1916, is produced naturally by human liver mast cells and basophil leukocytes, and inhibits coagulation primarily by accelerating the interaction between antithrombin and thrombin. Nanorobot exteriors can be "heparinized", and thereby rendered thromboresistant by immobilized heparin on all blood-contacting surfaces at ~monolayer surface concentration (e.g., 7-10 pmol/cm^2). Cellulose membranes coated with 3.6 pmol/cm^2 of endothelial-cell-surface heparan sulphate show complete inhibition of platelet adhesion. If satisfactory passive nonthrombogenic surfaces cannot be found, nanorobots might employ any of several active strategies to prevent iatrogenic coagulation.

During development and after physical injury, new blood vessels may originate from pre-existing blood vessels by angiogenesis or from endothelial cell precursors (angioblasts) by a process called vasculogenesis; both processes are mediated by paracrine growth factors. In certain circumstances, such as wound healing, post-ovulation capsule repair, and exercise-related capillary formation to support development of new skeletal muscle, angiogenesis is critical in the adult human body and its lack has been associated with chronic renal failure and other pathological conditions. Without the ability to incorporate vasculoid plates into new capillary vessels, masses of new non-envasculoided tissue would accumulate around the original (envasculoided) tissues over time, reducing the effectiveness of the appliance. In principle, vasculoid can support angiogenesis and vasculogenesis by extending itself into the newly-formed space. Watertight multicomponent metamorphic surfaces have been described by Freitas), but special techniques will be required to extend a new vascular branch while maintaining continuous watertightness. An angiogenesis event can be detected by monitoring concentrations of angiopoietin-2, VEGF, and other relevant factors. Conversely, detection of tumor angiogenesis factors could induce a vasculoid-mediated reduction in oxygen supply, increased local delivery of antibodies, refusal to deliver angiogenesis factors, and optional intervention such as delivery of chemotherapeutic drugs or external signaling to call for targeted therapeutic intervention. It is envisioned that self-repair capabilities via active replacement of damaged components (from onboard inventories of spare parts including spare plates will be an important feature of the complete vasculoid system design. These capabilities may be extended to support angiogenesis. When angiogenesis is not desired (for example, near tumors), local concentrations of angiogenic factors can be regulated by the vasculoid appliance to minimize vessel formation, and the vasculoid can refuse to deliver nutrients to the undesired tissues even if ersatz capillaries form. Attention must also be paid to the removal and disposal (or temporary onboard warehousing) of damaged plates. In rare cases, traumatic events may damage plates (rather than simply causing them to separate temporarily), and radiation damage may cause rare failures. A capillary damaged beyond repair (e.g., in a severe contusion) may be abandoned and dismantled by the body; in such a case the vasculoid should cease delivery to that region and remove the plates. Damaged endothelial cells can be detected based on sensory data indicating localizing chemical imbalances, and new stem cells or endothelial precursor cells can be transported to the vicinity of the damaged site to help restore the natural vascular integrity. Such restoration in some cases may require repositioning of a few plates using the motive cilia.

One important function of fluid flow in blood vessels is to ensure vascular patency. But replacing bulk fluid with a vasculoid appliance is unlikely to cause the smaller vessels to collapse because the buckling force is increased by up to several orders of magnitude by the addition of the plates in the capillaries and related small vessels, even in the absence of bulk fluid. The largest vessels such as the aorta will lose significant luminal support (even if filled with gas), but—in addition to the external muscular and elastic support from the surrounding tunica media and tunica adventitia layers—such vessels also possess an extensive capillary vasculature which, greatly stiffened and at typical capillary densities, should provide adequate replacement support. Intravascular (cross-luminal) spring-loaded diamondoid scaffolding could also be added at need. Arterioles in the natural vascular system perform extensive regulatory functions. These and related functions (including vasoconstriction and vasodilation) can be controlled, in turn, by the vasculoid (e.g., by monitoring local NO concentrations). The underlying vascular musculature will probably be maintained in the relaxed state if this maximizes tissue stability and simplicity of control, although further research of this question is warranted.

Numerous local, intermediate-scale, and global control systems and protocols will be required to ensure the proper autonomic operation of the vasculoid appliance. Special communications subsystems for local plate configuration maintenance, cargo traffic control, and fast systemic response to large-scale external stimuli must be designed. An interplate packet-switching network for long-range communications, and to support real-time autogenous user control, should be considered. System control might be simplified using local computing centers throughout tissues, directing immune, angiogenetic, and trophic functions. Besides the interplate network, trans-tissue sonic or optical communication in principle could be employed between spatially close but topologically distant portions of the vasculoid network. Precise knowledge of the functions of various cytokines and other substances will enable the vasculoid to respond appropriately to sensor data reporting local extravascular concentrations of signaling molecules. Complexity of stimulation patterns must mirror physiological conditions as much as possible. Highly responsive and user-friendly graphic user interfaces may be useful to allow the patient to communicate quickly and easily with his or her appliance; several aspects of such interfaces have been described elsewhere, but considerably more work remains to be done. The computational requirements of directed container switching/routing by the ciliary system should be investigated further. A great number of software and computational architectural issues are as yet unresolved, including a detailed molecular routing logic for each substance to be transported, comprehensive details of nanorobot control protocols, container traffic flow patterns (e.g. flow lensing, lane definition, congestion waves and gridlock, flow viscosity, etc.), and software requirements including software complexity for large-scale control issues, a discussion of which are beyond the scope of this paper. (For example, it might be interesting to apply existing vehicular traffic simulations to transport in fractal structures.) Only since the mid-1990s have multirobot control issues begun to be seriously addressed by the broader research and technical community. Computation for vasculoid accident recovery and repair will be considerably more intensive than computation for maintenance or installation, and must also be deferred to another paper. However, a few basic observations may be made here regarding vasculoid materials transport computation requirements. Concentrations of water and glucose are normally maintained by the appliance near typical serum levels, using simple feedback loops driven by concentration sensors located on the tissue side of the vasculoid device. Oxygen and carbon dioxide are similarly controlled using sensors capable of measuring partial pressures, though the precise triggering thresholds may be modified in special circumstances (e.g. high altitudes, deep sea diving, hyperbaric chambers). A simple rate-control mechanism driven by sensor data compiled once every second, employing ~1000 distinguishable concentration levels requires ~10 bits/sec per molecular type, or ~40 bits/sec for all four molecules. Most of the time, entire rotor banks are engaged using a single command. Ciliary motions, largely stereotypical, likely require a similarly small computational budget.

Establishing or maintaining appropriate concentrations of mixed-cargo molecules is far more computationally complex. These other molecules include minerals, vitamins, lipids and waste products, but the most numerous and difficult to coordinate are the blood proteins and sterols including control molecules such as insulin, leptin, gastrin, resistin, estrogen, or testosterone; synaptic transmitters and other neural biochemicals such as acetylcholine, adrenalin, dopamine, nitric oxide, or serotonin; cytokines and other signaling molecules such as follicle stimulating hormone, epidermal growth factor, or tumor necrosis factor; and many thousands of specific antibodies that reflect the body's unique learned recognition of specific antigens. (Embedded analytical units scattered throughout the vasculoid surface can test for rising concentrations of hitherto unrecognized nonmetabolizable molecules, and configure a set of programmable binding sites to rotor these molecules out of the tissues as well, though this should be a relatively rare occurrence.) Since the human genome contains ~40,000 genes, most of which may encode several proteins each, we assume there may be on the order of ~100,000 distinct biomolecules that must be recognized, monitored, transported, and regulated at the docking bays. This is likely a generous estimate, given that most of the proteins produced remain intracellular, but our conclusions do not depend sensitively upon the precise figure chosen. Upon receipt of any tanker, each docking bay scans the tanker manifest before offloading the contents. In the case of gas, water, and glucose tankers, the manifest is extremely brief, since only one or two molecules are involved. This content information can be encoded in as little as 15 bits. A precise molecule count of up to 3 x 1010 molecules per tanker requires an additional 35 bits, but in most cases a precise molecule will not be necessary and a range figure requiring fewer bits should suffice. At the maximum rate of one tanker unloading every 20 seconds, the manifest-reading bit rate is at most ~2.5 bits/sec. In the case of mixed-cargo tankers, 17 bits uniquely specify the identity of each of 100,000 distinct molecules; each tanker can hold a total mixed-cargo molecule count of ~109, requiring up to 30 bits per item to ensure an exact count. Based on existing human physiology, we posit a specification that each vascular cell must have access to at least 5 per cent of all ~100,000 molecular types during one blood circulation time, ~60 sec. Since each cell is supplied by two mixed-cargo docking bays, then restricting tankers to a maximum of 834 different molecules per load fulfills our specification, requiring in the worst case a 39,198 bit manifest to be read in 20 seconds, a maximum bit rate of ~2,000 bits/sec. Again assuming ~10 bits/sec per molecular type for rate-control of rotor mechanisms, rotor control will require a maximum ~8,340 bits/sec. (One design alternative is to allow stochastic mixing and transport of the low-concentration molecules which would mimic the natural process, reduce the computational needs, and impose only a very small burden on the transport capacity.) In sum, docking bay process control appears to require ~10,000 bits/sec of computation. Cellulock control requires reading manifests of up to ~1000 cells of at most ~1000 different types, which is ~20 bits/cell type or ~20,000 bits/manifest. Cellulocks unloading one boxcar every 60 seconds will require a bit rate of ~333 bits/sec. Other control tasks will add only fractionally to the ~10,000 ops/sec budget previously estimated for individual docking bays, a capacity that can be used for other purposes between docking events.

A detailed assessment of vasculoid system reliability also lies beyond the scope of this paper, especially given the possibility for prompt and severe damage if the robot is subjected to extreme accelerative loads or unplanned segmentations. However, if we assume normal gravitational loading and ignore catastrophic accidents, we find that the vasculoid promises an annual survival probability well in excess of "six nines" (failure probability < 10-6) at least against radiation damage if all major subsystems incorporate tenfold redundancy. Adopting Drexler's radiation damage model for the first part of this analysis and applying it to a system comprised of N components, with each component comprised of n redundant parts, the probability P that the system remains operational after T years is approximated by: P = exp [-N (1—p)n], where p = the probability that an individual part remains operational; p ~ exp(-1015 Dm), where m = mass of the part in kg and D = radiation dose in rads ~ 0.5 T for normal background radiation in the terrestrial environment. As a further simplifying assumption for this analysis, we shall presume that all vasculoid modular "parts" are similar in mass to the ~4,000,000-atom robotic manipulator device, after which the vasculoid cilium is patterned. From the above formulae, p ~ 0.999960 for each such cilium-like part, per year, even with no internal redundancy within the part. The vasculoid ciliary subsystem includes a minimum requirement of N = 300 trillion cilia incorporating a redundancy of n = 10, giving a total of 3000 trillion cilia. From the above, the annual probability of failure of the ciliary subsystem is < 10-29; assuming a redundancy of only n = 5, failure probability is still < 10-7. Each vasculoid plate is ~2 micron^3 in volume. Given that there are ~150 trillion plates, with zero redundancy among them, and then to be assured of a complete plate subsystem annual failure rate < 10-6, the annual probability of failure per plate must be reduced to ~10-20. This is achieved for plates comprised of a total of ~88,000 "parts" as described above, with a redundancy of ~6 among such parts. However, in this study we have specified a more generous (i.e., safer) redundancy of 10 among such parts. Each vasculocyte repair robot is ~10-14 kg in mass. Given that there are ~200 billion active vasculocytes, and assuming zero redundancy among them, then to be assured of a vasculocyte subsystem annual failure rate <10-6, the annual probability of failure per robot must be reduced to ~10-17. This is achieved for robots comprised of a total of ~125,000 "parts" as described above, with a redundancy of ~5 among such parts; we have again assumed a more generous redundancy of 10 among such parts. Non-radiation failure mechanisms have not been examined in the above analysis.

Software failure modes have been ignored, and even plate computers may be complex enough to host intentionally disruptive programs such as computer viruses. Many other more subtle or second-order difficulties also have been ignored here, such as the potentially destabilizing effects of dynamic oscillations and resonances among the various moving components which would require a detailed design before undertaking a comprehensive analysis. For instance, we should try to avoid Per Bak's notion of self-organized criticality when complex systems are pushed too close to their design limits, as in car traffic simulations. Numerous specific failure modes have not yet been exhaustively analyzed, including complete single-plate failures and the seizing or locking of individual cilia while a transfer operation is in progress, e.g., near a capillary entrance where such failures could prove most troublesome. However, the vasculocyte fleet has been scaled to accommodate a reasonable repair mission requirement.

Vasculoid tankers are micron-size compressed-gas containers which may rupture explosively—though this event is extremely improbable, since tankers with rupture strength exceeding 40,000 atm are loaded to only 1000 atm pressure. If this unlikely rupture event occurs in a large vessel such as the aorta, the possible failure modes may resemble those involved in a detachment avalanche, but the low tanker mass should preclude any serious direct infrastructural damage. However, if the explosive event occurs within a smaller vessel such as a capillary, the risk of breach is considerably greater. However, the maximum overpressure that can be explosively applied to the appliance walls is limited to the ~1000 atm compression pressure of the tanker contents, which likely does not exceed the theoretical rupture strength of the vasculoid walls in capillaries which may be crudely estimated as pmax ~ 3300-33,000 atm, taking vasculoid plate tube thickness twall = 1 micron, capillary tube radius R ~ 3 microns and working stress sw ~ 1010 N/m^2 for solid tube walls or maximum interbumper working stress sw ~ 109 N/m^2 for more realistic plated walls jointed with bumper interconnects. Whether such explosive events will occur with sufficient frequency to warrant explicit corrective protocols (beyond conventional vasculoid self-repair activities) deserves further study. A shockwave cascade failure (such as recently occurred at the Super Kamiokande neutrino detector facility, wherein the initial implosion of a single photomultiplier tube (PMT) during refilling of the water tank triggered a runaway cascade, destroying 6,665 of the 11,146 PMTs in a few seconds seems unlikely in the vasculoid but also should be investigated theoretically.

In practice, one should expect that a few contaminant molecules will make their way into the vasculoid, perhaps from local strain that exceeds the adjustment capabilities of the metamorphic bumpers, accidental spills, and so forth. A molecule that is small enough to be mobile at room temperature is small enough to be captured by a sorting rotor. Larger molecules can be physically handled by vasculocytes. A specialized class of vasculocyte, also measuring several microns in size and moving at ~1 cm/sec, can employ a 0.25-mm-wide chemotactic detector bank on each mobile nanorobot to search for, bind, and remove large molecules—a complete sweep of the entire vasculoid luminal surface once a minute requires the activity of only 1 per cent of the total active vasculocyte fleet. The motion of the tankers creates a net motion of the nitrogen atmosphere that can sweep molecules toward centrally placed filters. Thus a combination of internal sorting rotors, filters, and vasculocyte activity can quickly extract contaminants from the vasculoid volume. Once contaminants are trapped, they may be dealt with in ways suggested previously. The volumetric capacity of the appliance to deal with internal contamination must be scaled according to anticipated applications and event scenarios.

Installation of the vasculoid involves, at the least, complete exsanguination of a sedated patient and an intricate vascular plating operation. Two hypothetical installation scenarios are presented. The first scenario is a detailed description of a procedure that would be feasible using the technology available currently. This is to demonstrate that vasculoid installation can in principle be carried out without violating any well-established medical or physical limits. However, the authors are aware that by the time a vasculoid-class device can be built, medical technology will have advanced significantly. We therefore also briefly sketch out a highly speculative second scenario which, if practicable in some future era, might be considerably more convenient and up to 100 times faster. This second procedure would be unduly aggressive by today's standards but might be feasible and safe, given the supporting technology available in a nanotechnology-rich medical environment. The principal installation scenario involves complete exsanguination of a sedated and hypothermic patient, replacement of the natural circulatory fluid with various installation fluids, followed by mechanical vascular plating, defluidization, and finally activation of the vasculoid and rewarming of the patient. Installation takes ~6.5 hours from start to finish and requires a peak ~200-watt power draw midway through the procedure. (By comparison, present-day kidney dialysis treatments require 4-12 hours and the equipment also draws a few hundred watts.) Our discussion first considers the lifespan of cells temporarily denied access to external molecular transport mechanisms. We next describe the details of vasculoid installation including patient preparation (~24 hours), vascular washout (~4 hours), vascular plating (~1 hour), defluidization (~0.3 hour), and initialization and cold start (~1.2 hours). We conclude with a brief discussion of vasculoid removal. The hypothetical installation protocol described below has been selected for maximum comfort, reversibility, and reliability. The correct performance of the system is verified at every step, providing a plateau of safe operation before moving on to the next step. Since the entire procedure is intended to be fully reversible at each step, any step which fails or does not proceed in a manner acceptable to the installing physician may be quickly abandoned with a retreat to the previous plateau of established safety, after which a decision may be made to try again or to abort the installation. Patients must be made aware that they are about to undergo a major medical procedure which involves replacing ~8 per cent of their body mass with complex nanomachinery. They must be psychologically prepared to deal with the personal implications of this. From the turn-of-the-century perspective, vasculoid installation appears massively intrusive especially when compared to other superficially related procedures with which we are commonly familiar such as intubation, kidney dialysis, blood transfusion and blood replacement therapies, installation of pacemakers or artificial organs, and coronary angioplasty or fiberoptic endoscopy. However, in a future era when nanomedical systems are widely employed and generally accepted as standard treatment, vasculoid installation may be regarded with considerably less trepidation than it would today.

In classical medicine, "ischemia" refers to a local inadequate blood supply which is usually caused by a mechanical arterial obstruction (e.g. clot), a spasm wherein a blood vessel pinches shut, arterial narrowing (e.g. arteriosclerosis), or cessation of cardiac activity. The principal clinical outcome of local ischemia is a shortage of oxygen. Because of the high continuous biological power density of nerve cells, neurons and the brain succumb most rapidly to oxygen starvation. The precise survival limit of human brain tissue under hypoxic conditions (after which time some decline begins to occur) depends on many factors. Traditionally this has been reported as ~4 minutes (240 sec), although more recently it has been shown that 10 minutes of warm ischemia (even followed by another 10 minutes of trickle-flow CPR) is reversible without neurological deficit if followed by mild hypothermia when normal blood flow is restarted, and even modest advances in technology will likely extend this still further. Much longer periods of cold ischemia are tolerable—for example, survival with complete functional and histologic cerebral recovery has been achieved with brain temperature at 5-10°C during 1 hour of circulatory arrest, and stable spontaneous circulation has restored after water ice submersion of up to 90 minutes. In nanomedicine, where it is possible to achieve precise control of human physiology at the cellular and molecular levels, ischemia may refer to an interruption in the extracellular mass transport of any vital molecular species within the human body. Failure to adequately import a particular nutrient, or to adequately export a specific harmful waste product, or to properly regulate the transport of a particular signaling molecule, may produce some of the classical symptoms of cellular ischemia. Since the entire bloodflow must be interrupted during the vasculoid installation process (producing whole-body ischemia), it is useful to estimate how long tissue cells can maintain normal metabolism and avoid toxemia after cellular access to the bloodflow is denied. This time period—the ischemispecific limit—varies with each important cytometabolic molecule, as summarized below. (The limit also varies by cell type; a more detailed study would likely reveal specific tissues where ischemic sensitivity is higher than average for a given metabolite.)

Oxygen—The average (20 micron)^3 tissue cell consumes ~107 molecules/sec of O_2 at the basal (resting) rate of ~30 picowatts. The cytosol of such a cell can dissolve up to ~6 x 108 O_2 molecules at 310 K. If suddenly cut off from all external supply, the average cell has only enough O_2 in inventory to survive ~60 sec at the basal metabolic rate.

Glucose—Assuming ~50 per cent energy conversion efficiency and ~10-3 gm/cm3 glucose in the cytosol, then ~3 x 1010 glucose molecules are available in the average tissue cell. For a power demand of 30 picowatts, ~4800 zJ/glucose molecule, and ~50 per cent energy conversion efficiency, then the basal glucose consumption rate is 1.25 x 107 glucose molecules/sec and the basal cellular ischemispecific limit for glucose is ~2400 sec.

Carbon Dioxide—The metabolic "combustion" of one molecule of glucose with six molecules of O_2 produces six molecules of CO_2, hence the average tissue cell produces ~107 molecules/sec of CO_2 at the basal rate. The cytosol of an average cell should normally dissolve up to ~1010 molecules of CO_2 at 310 K (assuming working tissue PCO2 = 54 mmHg and applying the Henry's law constant for CO_2, an estimate which compares favorably with the measured ~5 millimoles/kg intracellular CO_2 concentration (2.4 x 1010 molecules/cell assuming an 8000 micron3 cell) in various frog tissue cells at 302 K. Hence the cellular ischemispecific limit for CO_2 is ~1000 sec, consistent with maximum physiological cellular and plasma pH levels. It may be possible to dissolve more CO_2 but this will lower pH, leading to toxic acidosis of the cell. (Acidosis has two major components, CO_2 and lactic acid. The absence of oxygen blocks electron transport and stalls the Krebs cycle, halting CO_2 production, so glucose is shunted to lactic acid in anoxia.) Additional amounts of CO_2 may be stored in combination with plasma proteins as carbamate.

Lactate—Under anaerobic conditions, lactic acid is produced during the metabolism of glucose (glycolysis) in most cells, glycogenolysis in muscle and glucolysis; two molecules of lactic acid are produced per molecule of glucose metabolized. Thus at the basal rate the average anaerobic tissue cell can produce at most 2.5 × 107 lactate molecules/sec—though this is an overestimate by perhaps a factor of 2 because in most cells the pyruvate (the immediate precursor of lactic acid) formed at the end of glycolysis enters the TCA cycle and is further oxidized by mitochondria. The maximum normal blood concentration of lactate equates to ~1.1 × 1010 molecules/cell, potassium channel openings are induced by 2-20 nM lactate (0.96-9.6 × 1010 molecules/cell) applied to the cytosol of rabbit ventricular myocytes, and the brain lactate threshold for cerebral ischemic damage is 17 mmol/gm or ~8.1 x 1010 molecules/cell, so the ischemispecific range for lactate is ~380-3800 sec.

Nitrogenous Waste Products—Urea is the principal mammalian waste product due to protein, purine, and pyrimidine nitrogen metabolism, constituting ~85 per cent of human nitrogen excretion. However, urea is formed in the human liver through reactions of the Krebs ornithine cycle, a process not available to tissue cells denied access to the circulation. Ammonia is the chief byproduct of protein and amino acid metabolism in the cell, with one molecule of ammonia produced per molecule of amino acid broken down. The daily RDA for protein represents an upper limit on the breakdown rate of ~5 × 105 amino acid molecules per cell-sec, assuming mean amino acid MW ~ 100 Da, implying a maximum ammonia generation rate of ~5 × 105 molecules/cell-sec. Given the maximum observed concentration of ammonia of ~6 × 108 NH3 molecules/cell in whole blood cells, the ischemispecific limit for nitrogenous wastes is ~1200 sec.

Acetate—Acetic acid (MW = 60 Da) is the simplest possible fatty acid having an even number of carbon atoms. It is most notably produced during the breakdown of acetaldehyde (the second step in the metabolism of alcohol), bacterial fermentation in the gut, and in other circumstances. The typical human intracellular production of acetic acid has been roughly estimated from rat studies as ~8 x 106 molecules/cell-sec. This chemical is highly metabolizable in vivo, but assuming a maximum nontoxic limit of ~10-4 gm/cm3 (vs. ~4 × 10-3 gm/cm3 for all esterified fatty acids in cells) or ~1010 molecules/cell, then the ischemispecific limit for acetic acid is ~1300 sec.

Ketones—Absent circulatory removal and in cases of low glucose levels, the metabolism of fats may produce abnormal amounts of toxic ketones including primarily beta-hydroxybutyric and acetoacetic acids (and their decarboxylation product acetone) which may be present up to ~6 × 108 molecules/cell assuming MW ~ 86 Da for ketones. Given the RDA for lipids of ~7 × 10-7 kg/sec and since one palmitic acid (typical lipid) molecule (MW = 256 Da) would convert to ~3 molecules of acetoacetic acid (MW = 86 Da), the maximum natural ketone generation rate is ~5 × 105 molecules/cell-sec and the minimum ischemispecific limit for ketones is >1200 sec.

Summary of Cellular Ischemispecific Limits—If the average human tissue cell is denied access to all extracellular molecular transport systems, the major byproducts of normal cellular metabolism may build to near-toxic levels from a near-zero initial concentration in ~103 sec. Glucose reserves last >103 sec, but oxygen runs short in ~102 sec without external resupply. These theoretical limits appear crudely consistent with the concept of the "Golden Hour" in traditional trauma care. For instance, ischemia induced by aortic cross-clamping in dogs has been survived for up to 20-60 minutes without inducing paraplegia (though decline of spinal cord electrical function is detected in the ~20-30 minute range for dogs and rabbits). Cardiac ischemia induced by aortic cross-clamping in dogs is survived for 30 minutes during normothermic (37°C) cardioplegia, 45 minutes during mild hypothermic (28-30°C) cardioplegia, and 90 minutes using potassium verapamil during profound hypothermic cardioplegia at 8-10°C. The duration of cold ischemic tolerance is enhanced by the administration of excess insulin and other pretreatments, possibly involving gene products that will be well-known in a future nanomedical era in which vasculoid installation is commonplace. (Interestingly, wood frogs can endure freezing for >2 weeks with no breathing, no heart beat or blood circulation, and with up to 65 per cent of their total body water as ice)

Patient Preparation (~24 hours): Before the vasculoid may be installed, the patient must be prepared as follows:

1. *Vascular Conditioning*—The day before the installation, the patient receives an injection containing a treatment dosage (~1 cm3) of vascular repair nanorobots or ~70 billion individual devices. These 8-picogram, 7 micron3 mobile legged artery-walking nanodevices clean out all fatty streaks, plaque deposits, complex atherosclerotic lesions, infections, vascular wall tumors, and parasites, and repair all other vascular lesions as required in less than 24 hours, in a multistep process described elsewhere in detail. As one additional task, these nanorobots inject each of the ~1012 endothelial cells lining the human vascular tree (a) with cytokine blockers to at least partially inhibit the cells' natural surface pressure and shear force responses, and (b) with adhesion-inducing glycoproteins to forestall cell migration during the installation procedure by encouraging firm anchoring. All vascular repair devices are then exfused before vasculoid installation begins; the results of their vascular reconnoiter protocol may be downloaded

to an external computer and used to prepare a detailed map of the patient's vascular tree to improve efficiency during plating and plate initialization. The patient may be started on a diuretic (e.g. furosemide) to reduce blood volume a few hours before the installation procedure begins.

2. *Sedation*— On the day of the operation, the patient arrives at the installation facility and is administered a preoperative sedative such as sodium pentobarbital an hour before the procedure is to begin, to encourage drowsiness.

3. *Cannulation*— A standard closed-chest cardiopulmonary bypass (aka. heart-lung machine support) or CPB is employed. Usually this involves a combination of femoral and thoracic vessel cannulation, but in this case a fem-fem bypass (cannulation of the vena cava and aorta via the femoral vein and artery) might suffice. This procedure, by definition, continuously supplies the equivalent of resting cardiac output through the femoral vessels and the blood pump/oxygenator circuit—typically a 50-70 cm3/sec flow rate during hypothermic CPB, allowing the entire human blood volume can be exchanged about once every few minutes. Fresh oxygenated blood flows retrograde up the femoral artery into the aorta, and oxygen-depleted blood flows retrograde down the vena cava to return to the heart lung machine via the femoral vein. The leg inferior to the cannulation point is supplied by collateral vessels. At this point in the procedure, the patient's own blood continues to circulate through the body. (Note that the above "manual cannulation" scenario is included for illustrative purposes only. In an era when advanced medical nanotechnology is available, self-directing nanocannulators will make it easy to quickly and safely establish flow-regulated channels into any desired artery or vein, up to and including the aorta and vena cava, so much higher flow rates than the conservative figures used here are theoretically available if required.)

4. *Heparinization*—Heparin and streptokinase are injected (a) to prevent clotting and promote lysis of hemostatic fibrin, (b) to break up axial red cell rouleaux that form spontaneously at low blood flow shear rates, and (c) to help ensure patency of the catheters throughout the procedure. The conventional cocktail of drugs given to patients placed on bypass leading up to a cold circulatory arrest procedure is rather complicated and will not be elaborated further here, but there are established procedures in conventional medical practice for cooling patients and diluting their blood for work at very cold temperatures.

5. *Cytoactivity Restraint*—The installing physician next administers: (a) an

erythropoietin antagonist to maximally suppress erythrocyte production; (b) agents to temporarily suppress all leukocyte production, reversibly depress leukocyte intracellular metabolism and cytokine sensitivity, and briefly reduce or block antigen reactivity; (c) agents to temporarily suppress platelet production, reversibly depress platelet intracellular metabolism and cytokine sensitivity, and temporarily toughen the platelet cytomembrane to reduce the likelihood of shear damage; (d) minute traces of normal bloodstream hormones, cytokines, and other control biochemicals designed to minimize production of natural secretions such as insulin, adrenalin and testosterone, glucose and cholesterol, sebum and semen, etc., as well as angiogenesis inhibitors to halt the development of new capillaries; and (e) broad-spectrum antibiotics or programmable nanobiotics such as microbivores designed to prevent microbial attack during the brief period of immunological vulnerability. For example, one element of this complex and as yet incompletely specified biochemical cocktail might be dexamethasone, the most powerful anti-inflammatory and immunosuppressive adrenocortical steroid; another element might be anti-CD18, an antibody known to prevent leukocytes from sticking to blood vessel walls in the brain. (Some of these and related restraints may not be absolutely essential, since the processes in question are quite slow compared to the time course of installation.)

Vascular washout (to prepare the patient for vasculoid plating) may be performed as follows:

1. *General Anesthesia*—A general anesthetic such as propofol is administered by the installing physician at sufficient dosage (~90 mg) to establish a condition of surgical anesthesia.
2. *Respirocyte Infusion*—Over a period of 3 hours, the patient's entire blood volume is replaced with a suspension of fully charged respirocytes (1 micron^3 spherical O_2/CO_2 1000-atm pressure vessels) in an isotonic aqueous 0.01 M glucose solution (double the natural serum concentration) also containing most of the cytoactivity restraint substances described earlier and additionally a mixture of appropriate electrolytes and other components commonly found in hypothermal blood substitutes. A 5 per cent (by volume) respirocyte suspension, containing ~1011 respirocytes/cm^3, is sufficient to provide oxygen and carbon dioxide transport equivalent to the entire human red cell mass for ~104 sec (almost 3 hours) after the cessation of respiration. The respirocyte fleet generates ~17 watts of waste heat while supplying oxygen at the human basal rate, producing at most a negligible ~0.1 oF rise in core body temperature. Typically ~20

liters of infusant is required to completely clear the vessels of all natural blood components. Exfused blood cells are removed, collected and saved for emergency reperfusion, if required.

It is true that the reduction of cerebral blood flow by 50 per cent can produce marked disturbances in brain metabolism, and at 20 per cent of normal flow the neurons can depolarize with rapid loss of intracellular potassium into extracellular spaces. But the infusant composition can be adjusted to maintain physiological electrolyte (esp. Na+, K+, and Ca++) concentrations and osmotic balances, and to extract any excitotoxins that might be released—or else respirocyte-class pharmacyte nanorobots can be employed to similar effect. Since blood cells have been removed, leukocyte plugging and other forms of neutrophil-related damage cannot occur. The basal metabolism remains sufficiently active throughout the procedure so that tissue cells suffer minimal ischemic damage. The infusant constitutes an adequate temporary replacement for the natural blood which is being removed. This is based on an established medical technology: It is well-known that the entire blood supply of a human being can be replaced with cell-free blood substitute at a low temperature, with the patient then reperfused with blood and recovered, as was first demonstrated clinically 30 years ago as a treatment modality for hepatic coma.

3. *Cooldown*—Once the entire blood volume has been completely exchanged with the respiro-infusant described in (2) in the unconscious patient on CPB (as in conventional medical settings), control of perfusate temperature will suffice to control body temperature. Core temperature follows the perfusate temperature fast and close, with any perfusate/core temperature difference decaying with a ~10 minute half-life. The heat capacity of the body is so large that the 200 watts of power dissipated by vasculoid installation over an hour or so would heat the patient by only a couple of degrees, which is tolerable. (By comparison, external cooling of a body in circulatory arrest is so inefficient that even placing the patient in an ice water bath would yield surface-to-core temperature differences decaying with a half-life of hours.) The patient's average core temperature is reduced from 310 K (37°C) to 280-290 K (7-17 °C) in ~1 hour, a cooldown rate of ~0.5°C/minute. This cooldown schedule is qualitatively similar to hypothermic schedules commonplace in current resuscitation medicine and hypothermic CPB procedures. Mivacurium, rocuronium, or older agents such as vecuronium or metubine (recently withdrawn) may be administered to inhibit shivering and as a reversible nondepolarizing muscle relaxant; the effect of vecuronium bromide may be reversed by

acetylcholinesterase inhibitors such as neostigmine or pyridostigmine.

B. Wowk notes that there are three broad categories of clinical hypothermia used in medicine. First, there is mild hypothermia (a few degrees below normal body temperature). This is very commonly used during cardiopulmonary bypass for open heart surgery, and often occurs as an incidental effect of general anesthesia during other surgeries. Second, there is deep hypothermia (15-20°C). This is used when significant periods of circulatory arrest (up to 1 hour) are required for complex bloodless surgeries such as repair of cerebral aneurisms or congenital defects of the aortic arch. Hemodilution (partial blood substitution) is used during these surgeries, in part to avoid red blood cell agglutination at low temperatures. Third, there is profound (or ultraprofound) hypothermia (0-10°C). Profound hypothermia does not yet have routine clinical use but is the subject of active investigation in relevant animal models with an eye toward human use, and preliminary human clinical results recently have been reported. Total blood substitution, as proposed in the vasculoid installation protocol, is the norm for profound hypothermia. Specialized perfusates such as hypothermosol are necessary to overcome physiological problems with this temperature regime that ordinary plasma can't handle. Dogs have been held for three hours in profound hypothermia during perfusion of a blood substitute, and the record for profound hypothermic perfusion without neurological deficit is said to be ~6 hours; the record for profound hypothermic circulatory arrest (as distinct from continuous perfusion) in dogs at <5°C is ~3 hours.

4. *Cardioplegia*—After the cooldown process has begun, the heart (already in bradycardia from the muscle relaxant) may be stopped by direct infusion of cold potassium chloride solution (reversible with atropine or digitalis) or other standard cardioplegic solution. However, if the target is ultraprofound hypothermia there is no need to stop the heart with drugs during this process because the heart will stop itself when the temperature falls below 15-20 °C.

In ~4 hours, washout, cooldown and cardioplegia are completed. Cytometabolic processes continue, but, at 280 K, with reduced metabolic requirements and hence reduced oxygen demand. This potentially allows infused respirocytes to provide complete respiratory gas maintenance for up to ~105 sec (~1 day) before they would be exhausted after complete cessation of circulatory flow via the catheters, giving a significant safety margin for the next phase of the installation procedure. However, B. Wowk notes other risks of profound hypothermic circulatory arrest beyond mere hypoxia, including the failure of ion pumps (causing growing intra- and extracellular imbalances of calcium ions), the alteration of cell membrane permeability (allowing normally extracellular solutes to slowly leak into cells), and the decoupling of certain cellular metabolic processes (which does not occur during mild hypothermia). The vasculoid installation process may fit within the circulatory arrest times achievable with the standard off-the-shelf deep hypothermic (not profound hypothermic) protocols of conventional medicine. Further study is required to determine the optimum temperature for hypothermic installation.

The anesthetized patient is finally ready for intravenous deployment of vasculoid components.

The original 5 per cent respirocyte suspension is replaced by a new suspension containing 1 per cent fully-charged respirocytes and 10 per cent cargo-bearing vasculocytes (plus cytoactivity inhibitors and glucose), creating a mixture whose viscosity and flow characteristics are roughly equal to normal-hematocrit human blood. At an ~11 per cent nanocrit, partial plug flow might ensue in a few of the smallest capillaries, but complete plug flow (requiring much higher pumping power and pressure than laminar flow) can be avoided. Each 3 micron^3 vasculocyte grasps a single 2 micron^3 plate, ready for installation, thus the new vasculo-infusant contains ~20 × 109 vasculocytes/cm^3. Installing ~150 × 1012 basic plates thus demands a minimum of 7500 cm^3 of vasculo-infusant, requiring ~3800 sec or ~1 hour assuming a very gentle flow rate of only ~2 cm^3/sec. Each vasculocyte drifts quietly in the flow until it encounters a vessel wall for the Nth time (N is an integer control variable), which activates it, causing it to attempt to release its cargo in a clear space. (Slowly increasing N during the installation process produces a crudely progressive plating pattern.) If the immediate area is already fully plated, the legged vasculocyte walks across the surface until it reaches a clear area to deposit its cargo. Corner registration of adjacent plates is verified prior to plate release, to ensure a maximally dense tiling pattern. Because of their small size, vasculocytes can enter even the narrowest capillaries and install plates by tiling motions. This process may be loosely regarded as a robotically-guided variant of fluidic self-assembly, a well-known existing commercial process. Special procedures must also be devised to handle encounters with nonendothelial materials (e.g., fibrin strands) or stray cells (e.g., adherent leukocytes) that may be blocking the endothelial surface, or denuded patches of vasculature lacking full endothelial cell coverage, although most of these defects should have been remedied during preoperative vascular conditioning. Once its cargo plate is in place, the vasculocyte releases back into the flowing fluid, powers down, and is eventually exfused from the body. Approximately 42 billion plates/sec are deposited during this 1-hour process. We generously assume that each 1-50 picowatt vasculocyte requires ~100 sec of active operation up to peak power to find a clear space to deposit its cargo before resuming its fluidborne dormancy—though early arrivers will spend less time searching for gaps in the structure than later arrivers. Thus there are at most ~4 trillion vasculocytes active at any moment and the total power released as waste heat by the vasculo-infusant is under ~200 watts, thermal energy which is easily carried off by the cold infusant. Given that one plate is installed by one vasculocyte carrier that performs ~106 mechanical motions/sec during a ~10 sec install time, this implies ~107 motions/ plate and allows an allocation of ~1800 motions per nanometer of plate perimeter during the installation of each plate, which seems sufficient. The plates are not passive during the installation process. Besides the 20 cilia positioned atop each plate to provide tanker and boxcar mobility as part of the ciliary subsystem, plates also possess "motive cilia" to assist in both installation and repair operations. Each motive cilium is fully retractable when not in use. Complete positional, rotational and translational control during installation (and functional redundancy) requires at least one motive cilium on each of the 4 sides and at least four on the underside to establish a stable tripod while walking. The motive cilia allow limited trans-endothelial cytoambulation by each plate (using adherent tool tips) and fine control of plate/plate jostling motions. A working cilium consumes 0.1 picowatt during continuous operation at 1 cm/sec roughly the speed of jostling motions involving motions of ~1 per cent plate width per microsecond. Even with all motive cilia operating at once, plate power is under ~1 picowatt; even with all 150 trillion jostling at once, maximum power would be under 150 watts—energy easily obtained from glucose and oxygen provided by the vasculo-infusant medium, and waste heat that is easily dissipated by thermal conduction. However, the motive ciliary power draw of the entire plate population normally should not exceed ~1 watt during installation because once properly positioned in a regular grid pattern (in <1 sec) all local plate jostling ceases and the motive cilia are stowed. The installation process may be made more efficient if plates are programed to dynamically shift holes (in the plating pattern) upstream. This ensures that gaps are unlikely to persist in the developing vasculoid structure (except in cases of plate malfunction), and also ensures that shifted holes will arrive in a known region of the upper vascular tree where they are most convenient for vasculocytes to quickly find and fill. Out-of-register plate domains that collide during this dynamic repositioning process require a simple interaction protocol to allow them to mutually align correctly (such a protocol has not yet been devised). Once plates are in place, their surface cilia can be adjusted to provide variable drag on the circulating fluid. This can be used as a crude method of directing resources for more efficient construction. Plates also assist the vasculocytes in system validation. Upon installation, plates briefly activate all subsystems to verify that everything is working properly. With all 18,750 sorting rotors spinning, individual plates momentarily draw up to ~2 picowatts. Malfunctioning plates jettison themselves back into the fluid flow for ultimate removal, or if this is not possible, take other action to bring their condition to the attention of the circulating vasculocyte fleet so that they may be replaced at once. Following supravascular positioning and subsystem validation, each plate inflates fluidtight metamorphic bumpers along its contact perimeter with its neighbors. This provides at least ~14 per cent linear effective elasticity (1.6 microns vs. 1.4 microns, center-to-center, comparable to the usual stretch requirements of the elastic arteries during normal cardiac pumping) and permits plates to track lateral movements of the underlying tissue while avoiding any relative movement of the opposing surfaces. Ventral lipophilic anchors dropped into the lipid bilayer cytomembrane of the adjacent endothelial cells of the vascular wall serve as sensors to detect any such relative movement, providing continuous feedback to control bumper inflation—although with cardioplegia such wall movements will be quite small because infusant pressure may be held nearly constant throughout the installation procedure. Neighboring plates lock their bumpers firmly together with reversible fasteners embedded in the bumpers. Along vessels whose diameters change rapidly as a function of axial position, orthogonal plate alignment along a circumference can be maintained via bumper expansion. Once maximum bumper extension is reached, a further increase in vessel circumference is accommodated in the next row by deflating bumpers in the circumferential direction and inserting one additional plate. Incommensurate offsets are reconciled using multiple ports in the bumper structure. If bumpers can expand by up to 20-30 per cent, then the blood vessel being plated can change in width by 20-30 per cent before such a pattern discontinuity is required, so such discontinuities should be relatively uncommon in most of the plated vasculature. Computer modeling of the dynamic adjustments required to achieve continuous plating at blood vessel bifurcations and in tapered vessels would be useful. As the natural vascular surface is covered by the growing vasculoid, individual plates that have completed self-testing procedures and are locked in place activate a sufficient number of rotors to begin providing full molecular transport services to the underlying tissues. Any plate malfunctions are corrected by the circulating vasculocytes, via plate changeout. Metabolism in the underlying cellular tissues continues normally, just as it did before these tissues were plated with the vasculoid. Rotors on the ventral plate face (away from the lumen) gain access to nutrients in the vasculo-infusant fluid via gated channels leading to the dorsal plate face (toward the lumen) which is directly exposed to the fluid; cellular waste products escape to the fluid by a similar route. Plate power requirements normally range from ~0.004 picowatt (basal rate) to ~0.09 picowatt (peak rate) plus ~0.6 pW for computation. At 280 K the patient's cells would require only a fraction of the basal rate; after plating is complete and in the absence of significant computation, the entire population of 150 trillion installed plates might generate only ~1 watt. Specific procedures for cellulock installation have not been examined in detail because these vasculoid components are relatively few in number and require a net installation rate of only ~9 million/sec (compared to 42 billion/sec for basic plates). To the extent cellulock placement can be governed by easily detected geometric factors such as vascular diameter (e.g., near capillary entrances), cellulock placement may occur similarly to plate deposition, but using vasculocytes programed with specific geometric release criteria. Some modest number of cellulocks destined for specific vascular addresses may be installed after plate initialization using a combination of vasculocytes and now-active cargo cilia, by replacing temporary plates with cellulocks. Nontubular vascular segments including flaps, valves, sphincters, portals, sinuses, plexuses, nodes, and discontinuous microvascular beds such as the red pulp of the spleen constitute only a tiny fraction of the vascular surface but may require special procedures by the vasculocytes. Endothelial pressure/shear responses may be managed by continuous emission of appropriate blockers, inhibitors or cytokines; the response may be reversibly disabled or eliminated using a designed vector to insert a short segment of revised DNA into the cells' nucleus. The vasculoid also includes an explicit mechanism to inhibit the electrical output of the sinoatrial node, thus maintaining a permanent (but in theory reversible) state of cardioplegia. After ~1 hour, the structure of the vasculoid is almost complete. All major components have been tested and are in good working order—indeed, the still-submerged vasculoid is already maintaining full cell metabolism in the tissue below. The patient is now ready for defluidization.

At this stage, the 1-micron thick monolayer of nanorobotic plates forms a chemically inert, flexible sapphire liner on the luminal (interior) surface of the entire vascular tree. This liner may be made fluidtight and airtight to at least modest pressures. Vasculo-infusant fluid is purged from the body by introducing ~6 liters of oxygenated pure anhydrous acetone as a dehydrating rinse followed by pressurized dry air at 2 atm through the input catheters and allowing all lavage fluids to exit through the output catheters for ~15 minutes. If necessary, surface cilia may assist in the distribution of fluid or air flow to help ensure that no pockets of fluid remain trapped anywhere in the system, and to ensure that no respirocytes, inoperative or damaged vasculocytes, or other stray particles are left behind. The upper limit for laminar flow (Reynolds number ~ 2000) of aqueous vasculo-infusant fluid passing through a smooth 9-mm diameter exit catheter occurs at ~1 m/sec. Assuming a constant laminar outflow velocity of 1 m/sec, flow volume through each of the two cannula is ~60 cm3/sec and the vasculoid lumen is emptied of all installation and rinse fluids in ~100 sec. (This step is easily reversed by re-introducing an aqueous oxygen-rich or respirocyte-laden nutrient solution.) Acetone viscosity is ~40 per cent that of water, so pumping power may be reduced as the rinse proceeds. Oxygen needed to sustain human life is slightly restricted for only ~100 sec during the pure acetone rinse, and thereafter is continuously available at the requisite concentration from the dry purge gas. A temporary supply of dry air with 20 per cent O_2 at 2 atm puts 5×10^{22} O_2 molecules into the vasculoid lumen, which when burned with glucose at 50 per cent efficiency in tissue cells produces ~20,000 joules, equivalent to ~2000 sec of power at the reduced metabolic rate of ~10 watts in a human body cooled to 280 K. The plate population consumes only an additional ~1 watt. As long as fresh purge gas continues to circulate, the oxygen requirements of both human and vasculoid can be satisfied indefinitely. Cellular glucose reserves will last ~10^5 sec at the reduced metabolic rate; subject to cautions noted earlier, cellular waste products may not approach toxic levels for ~10^4 sec. Vasculoid plates can also continue functioning in the temporary absence of glucose. Assuming a power requirement of ~0.6 pW/plate at the basal rate, then a ~1 hour supply of glucose may be stored in a small fuel tank inside each plate, representing only ~9 per cent of the 2 micron^3 plate volume. The final steps of vasculoid installation include:

(1) *Plate Initialization.* To aid in injury response and medical diagnosis, each plate may be assigned a unique position and identity code which can be echoed upon interrogation. With 200 billion operational vasculocytes and 150 trillion plates to initialize, each active vasculocyte must contact and initialize ~750 plates. Traveling at a net velocity of 10 microns/sec across the vasculoid, a vasculocyte can traverse (and initialize) 750 plates in ~100 sec. The address block is 100 bits in length; 50 bits are sufficient to specify 2^{50} = 1000 trillion unique objects in the system, but additional bits are required to specify each plate's branch level in the fractal vascular tree and 10 bits are reserved for self-correcting parity checks.

(2) *Install Storage Vesicles.* The vesicles are small storage garages containing reserves of mobile and cargo-carrying nanodevices, other auxiliary nanodevices, spare parts, replacement modules and repair materials, critical biochemical consumables, and compressed refuse. Vesicles may also incorporate nanoscale bioprocess plants to manufacture small quantities of artificial biochemicals not normally produced in the body. Each vesicle is ~1 mm^3 in volume, providing enough space to store ~1 billion tankers, ~300 million vasculocytes, or 350,000 boxcars. Vesicles may be attached directly to the vasculoid surface using sapphire struts affixed to reinforced base plates, or may be attached to each other to build convenient three-dimensional configurations (e.g. "bunch of grapes" formations).

As an order-of-magnitude estimate of the total number of vesicles that might be required, the volume required to store a complete replacement for the entire mobile container population would include 166.2 trillion tankers (1 $micron^3$/tanker) or 166.2 cm^3, 32 billion boxcars (2827 $micron^3$/boxcar) or 91.7 cm^3, and 2 trillion backup vasculocytes for tenfold redundancy (3 $micron^3$/vasculocytes) or 6 cm^3, totaling ~264 cm^3 of storage. Including additional space for bioprocess plants, spare parts, consumables and refuse storage may boost total vesicle volume to ~500 cm^3, requiring ~0.5 million vesicles which would fill roughly the combined internal volume of the four cardiac chambers which are no longer needed for pumping blood. However, conservative design principles would suggest that vesicles should be widely distributed throughout the body to maximize system survivability in the event of massive trauma. Infusion of 500 cm^3 through the two catheters at a modest infusion velocity of 1 cm/sec requires ~200 sec for infusion of all vesicle components and their contents. Each 3 mg (fully loaded) vesicle is installed by a team of ~3000 vasculocytes which can collectively apply ~30,000 nanonewtons of force (using only 10 active legs among the ~100 available on the ventral side of each vasculocyte) resulting in a vesicle acceleration of ~1 g (enough to overcome gravity in a supine patient, and easily increased if needed). Simultaneously installing all 500,000 vesicles requires the cooperative activity of ~1.5 billion 1-50 picowatt vasculocytes which consumes a total power of 1.5-75 milliwatts during the installation.

(3) *Install* sealed carotid, jugular, or navel access port, as required (~40 sec).

(4) *Activate* ciliary distribution system (~10 sec). This should probably include a brief stereotypical self-cleaning protocol to further ensure that all particles and stray nonessential nanorobots have been removed from the appliance interior.

(5) *Introduce* into the vasculoid interior the entire 257.9 cm^3 operational tanker and boxcar population, requiring ~100 sec at a modest ~1 cm/sec dry infusion velocity.

(6) *Warm* the patient (~1 hour) using heat generation from excess ciliary activation, energy from onboard thermogenerative systems or external heat sources imported via enhanced plate conductivity (or simple electrical or rf heating. Restart normal respiration as soon as possible. (The heart remains permanently inhibited, although the endocardium is coated with vasculoid plates.)

(7) *Reverse* cytoactivity inhibition that was initiated during patient preparation as appropriate.

(8) *Remove* catheters and seal all vascular and dermal breaches.

The vasculoid is now fully operational and self-contained. The patient is warm and breathing. All essential metabolic and immunological systems have returned to normal function. Healthy vascular conditioning is maintained on a permanent basis (post-installation) because the vasculoid exercises precise control over the transport of cholesterol, leukocytes and platelets, the principal participants in the arteriosclerogenic process. Skin and tissue suppleness may be controlled by adjusting the spring constants of linked bumpers between adjacent plates. A number of cosmetic issues must be addressed. For example, unpigmented tissues (e.g. the tongue and gums, fingernail dermae, the uvea and lacrimal apparatus of the eye, and brain tissue) will appear a pale, waxy translucent white owing to the complete absence of red cells; pigmented tissues (e.g. dark skin) should be largely unchanged. For more traditionalist users or for other aesthetic reasons, semi-natural coloration might be restored by retaining blue sapphire in venous channels but substituting red ruby (chemically similar to sapphire) in arterial plate materials. However, transmission and reflection properties of these substances may differ markedly, and such colors would likely require at least small numbers of potentially troublesome impurity atoms (10^{-3}-10^{-4} of Fe, Ti, or Cr atoms; to be present. Alternatively, a thin diamondoid-veneered sublayer of organic chromophores of appropriate colors might be employed. After a brief period of rest, postoperative checkout, and familiarization with user interfaces, the patient is released from the installation facility.

The procedures outlined above for vasculoid installation are designed to be fully reversible at every step. Thus, vasculoid removal may be accomplished by reversing the installation procedure, except that patient preparation should be performed first. If the patient has worn his or her vascular appliance for a considerable time, heart muscles, venous valves, splenic filtration systems and the erythroid marrow may have significantly atrophied, requiring remedial cellular repair. Homologous formed blood elements such as RBCs must also be premanufactured for postoperative infusion prior to vasculoid removal, and any genetically modified endothelial cells must be reprogramed back to their original state.

For our highly speculative second scenario, we briefly (and only superficially) describe an aggressive installation procedure that attempts to install the vasculoid into a fully metabolizing normothermic human body about 100 times faster than the hypothermic procedure detailed in the principal scenario. As discussed earlier vascular conditioning and mapping should occur before installation, in part to allow the manufacture of a folded, pre-assembled custom appliance tailored to the patient's unique vasculature which is then presented to the physician for installation. Our aggressive procedure involves three steps: (1) pre-charging the body with respirocyte infusant and removal of bloodborne cells; (2) physical installation of vasculoid plates as continuous rolling sheets, coincident with removal of infusant fluid; and (3) configuration and permanent connection of the appliance, followed by removal of support structures and system activation. It is intended that all steps will be carried out within the ~10^3 sec (~15 min) ischemic time limit imposed by non-respiratory metabolite concentrations. The aggressive procedure begins by accessing one or more large veins to permit rapid infusion and exfusion. The blood is quickly replaced, as it circulates, with an infusant solution containing electrolytes, glucose, and other essential substances, plus a sufficient nanocrit of respirocyte-class devices to provide respiratory support. The transfusion is speeded by using moderate numbers of "dragnet" style nanorobots as described in other contexts by Freitas—each such nanorobot manipulates an adjustable 1-100 micron diameter net in order to physically gather bloodborne cells that may be trapped in eddies near venous valves, vascular sinuses, and similar spaces. Within a few circulation times, perhaps ~200 sec, the vascular compartment is cleared of all free cells and most macromolecules. Exfused blood is retained until the procedure is complete, to facilitate reversal in the event of a medical emergency. The heart is now stopped. Respirocytes continue to provide oxygen and absorb carbon dioxide; even a modest 1 per cent respirocrit can provide ~25 minutes of resting metabolism without recharging, perhaps slightly less in the most energy-intensive tissues such as the brain. But because fresh infusant can be introduced continuously, the patient can be parked in this state indefinitely if need be. Next, the vasculoid is introduced into the arteriovenous vasculature directly through the heart via cardiopuncture or cardiocentesis. Cardiocentesis involves the surgical puncture of the heart, most often used today *in utero* on developing fetuses either to extract or to insert fluids into the vasculature (e.g., blood transfusion), or more rarely on adult organisms for related purposes. Cardiocentric installation of the vasculoid is required because the human circulation consists of two independent arterial circuits (pulmonary and systemic) each containing their own capillary beds. The appliance is installed as a continuously-everting concentric tube, a process called progressive fractal eversion that may be visualized as turning a glove inside out. Basic plates, docking bays and cellulocks are prefastened in the proper configuration to fit the various diameters and branchings of the blood vessels that they will coat. The necessary flexibility of this sheet of plates is provided by sophisticated watertight jointed interplate bumpers that have yet to be designed. Eversion may be powered by compressed gas, ciliary action between opposed plates, or by other appropriate means; the presence of pressurized gas (e.g., pure dry N_2) would also help to minimize the occurrence of leaks during installation. Each docking bay has an appropriate tanker already docked. As soon as the plate makes contact with the endothelial surface, the docking bay begins to work. Since the vasculoid contains a sufficient number of docking bays to accommodate peak metabolic loads, the preattached tankers can supply oxygen, glucose, and other nutrients, and accumulate wastes. Most critically, 5 trillion oxygen tankers are attached to all available respiratory-tanker docking bays, providing ~474 sec (~8 min) of oxygen at the basal rate. This sustainable duration is easily extended to ~1318 sec (~22 min) of basal oxygen supply by redesigning 13.9 trillion of the 24 trillion docking bays that are not needed for nonrespiratory transport to include the capacity for gas transfer during installation, since at the basal rate docking bays require only 67 active sorting rotors. Installation is initiated cardiocentrically as two primary segments imported via two cardiopuncture entry points through the relatively thin walls of the right atrium and the left atrium, respectively. Installation works outward from each chamber opening, usually in the natural direction of valve motion, moving towards the distal capillary beds. From the right atrial entry point, the unfolding right atrial appliance segment has four "fingers" and simultaneously plates:

(1) the superior vena cava; (2) the inferior vena cava; (3) the coronary sinus (which receives cardiac veins from heart tissue); and moves through (and plates) the right atrial and ventricular chambers, thence to plate; (4) the pulmonary artery. The right atrial segment also must plate all of the numerous Thebesian veins; these venules return blood from the myocardium without entering the venous current, and open directly into the right atrium. From the left atrial entry point, the unfolding left atrial appliance segment has five "fingers" and simultaneously plates: (1&2) the two left pulmonary veins (which frequently terminate by a common opening); (3&4) the two right pulmonary veins; and moves through (and plates) the left atrial and ventricular chambers, thence to plate (5) the aorta.

Additionally, a third vasculoid segment called the portal segment must be inserted through a third abdominal entry point in order most efficiently to plate the portal vein, which lies midway between the hepatic capillary beds and the intestinal capillary beds. Advancing at 1 cm/sec (2 cm/sec internal speed between the installed surface and the inverted tube), the vasculoid requires only 70 sec to plate the maximum ~70-cm main arterial or venous course. Fluid and respirocytes are withdrawn through the center of the inverted tube. As the vasculoid reaches the capillaries, the progression of the eversion may slow to ~10 micron/sec of travel, requiring another ~100 sec to complete all capillary plating. Human capillaries contain ~14 per cent of blood volume but comprise ~95 per cent of the surface area of the vascular system, so most of the appliance's plates are installed during this final ~100 seconds. If all 150 trillion plates are active throughout this process, the power draw is a physiologically-tolerable 150 watts. Upon first contacting the endothelium, each 2 mm^2 plate has ~0.1414 seconds—time enough for 14,140 localized 100-nm movements at 1 cm/sec (100 KHz)—to adjust its position and bumper configurations, and to clear away any detritus trapped beneath it (e.g., respirocytes, stray cells, etc.), before the next plate is placed. This seems sufficient.

Once the vasculoid has filled the capillaries, the arterial and venous branches are joined, detaching the internal fluid transfer channels running through the vasculoid lumen which are quickly retracted from the interior. The ciliary system can then be activated and tankers will begin to be delivered to and from the capillaries, starting ~200 sec after the first transcardial introduction of the vasculoid mechanism. Interior rinse is unnecessary because the luminal surface of the appliance was installed dry. Vesicle installation is performed post-installation via component injection, and assembled using vasculocytes, at leisure. Since this advanced model of the vasculoid comes preassembled, it can continually report its status and monitor physiological indications of the patient during installation, via an interplate communication network. In the event of an unforeseen medical catastrophe during installation, the vasculoid may be evacuated using an emergency extraction protocol if the appliance itself has not suffered a major loss of integrity and the transfer channels are still attached. In the emergency extraction protocol, the vasculoid undergoes partial flattening with circumferential fission in the capillaries. Then the arterial and venous sides of the vasculoid are withdrawn as they were inserted, using a reverse-eversion process with progressively increasing velocities possibly reaching ~10 cm/sec in the largest arteries. During such an extraction, only a ~1.4-micron-wide annular ring will be moving at this high speed adjacent to tissue—the installed vasculoid does not move, and the moving vasculoid is surrounded by immobile vasculoid. As a result, installed appliance plate-sheets may be pulled out by applying an extraction force to the fluid transfer channel structures that are still attached to the terminus of each capillary tube. Continuous pressurized return of respirocyte-charged infusant to the natural vasculature through these channels prevents pulling a vacuum behind the retreating tendrils of the appliance, which would cause the collapse of blood vessels as the appliance was removed. Thus, in less than a minute, the patient has returned to the relative safety of the respirocyte-infused parked state from which the installation procedure was begun. In the context of a medical technology capable of manufacturing a vasculoid, the temporary bloodless state would not appear to be a cause for concern.

REFERENCES

A. Carbone & N.C. Seeman, Molecular Tiling and DNA Self-Assembly, Aspects of Molecular Computing, Lecture Notes in Computer Science 2340, Springer-Verlag, Berlin, 61-83 (2004).

Akis, R., D. Ferry, C. Musgrave, "Kinetic Lattice Monte Carlo Simulations of Processes on the Silicon (100) Surface," *Physica E-Low Dimensional Systems and Nanostructures* 19, 183-187 (2003).

Albrecht, T.R., Akamine, S., Carver, T.E., and Quate, C.F. (1990) Microfabrication of cantilever styli for the atomic force microscope. J. Vac. Sci. Technol. A 8(4), 3386-3396

Albrecht, T.R., Grütter, P., Horne, D., and Rugar, D. (1991) Frequency modulation detection using high-Q cantilevers for enhanced force microscope sensitivity. J. Appl. Phys. 69(2), 668-673

Alexander, S., Hellemans, L., Marti, O., Schneir, J., Elings, V., Hansma, P.K., Longmiro, M., and Gurley, J. (1989) An atomic-resolution atomic-force microscope implemented using an optical lever. J. Appl. Phys. 65(1), 164-167

B. Ding, R. Sha & N.C. Seeman, Pseudohexagonal 2D DNA Crystals from Double Crossover Cohesion, J. Am. Chem. Soc. 126, 10230-10231 (2004).

B. Liu, N.B. Leontis and N.C. Seeman, Bulged 3-arm DNA Branched Junctions as Components for Nanoconstruction, Nanobiology 3, 177-188 (1994). N.C. Seeman, Molecular Craftwork with DNA, The Chemical Intelligencer 1(3), 38-47 (1995).

B.H. Robinson and N.C. Seeman, The Design of a Biochip: A Self-Assembling Molecular-Scale Memory Device. Protein Engineering 1, 295-300 (1987).

Binnig, G., Quate, C.F., and Gerber, Ch. (1986) Atomic force microscope. Phys. Rev. Lett. 56(9), 930-933

C. Mao, T. LaBean, J.H. Reif and N.C. Seeman, Logical Computation Using Algorithmic Self-Assembly of DNA Triple Crossover Molecules, Nature 407, 493-496 (2000).

C. Mao, W. Sun and N.C. Seeman, Construction of Borromean Rings from DNA, Nature 386, 137-138 (1997).

C. Mao, W. Sun and N.C. Seeman, Designed Two-Dimensional DNA Holliday Junction Arrays Visualized by Atomic Force Microscopy, Journal of the American Chemical Society 121, 5437-5443 (1999).

C. Mao, W. Sun, Z. Shen and N.C. Seeman, A DNA Nanomechanical Device Based on the B-Z Transition, Nature 397, 144-146 (1999).

E. Winfree, F. Liu, L. A. Wenzler, and N.C. Seeman, Design and Self-Assembly of Two-Dimensional DNA Crystals, Nature 394, 539-544 (1998).

E. Winfree, X. Yang and N.C. Seeman, Universal Computation via Self-assembly of DNA: Some Theory and Experiments, In: DNA Based Computers II, ed.by L.F. Landweber and E.B. Baum, Am. Math. Soc., Providence, pp. 191-213 (1998).

F. Liu, H. Wang and N.C. Seeman, Short Extensions to Sticky Ends for DNA Nanotechnology and DNA-Based Computation, Nanobiology 4, 257-262 (1999).

F. Liu, R. Sha and N.C. Seeman, Modifying the Surface Features of Two-Dimensional DNA Crystals, Journal of the American Chemical Society. 121, 917-922 (1999).

F. Mathieu, S. Liao, C. Mao, J. Kopatsch, T. Wang, and N.C. Seeman, Six-Helix Bundles Designed from DNA, NanoLetters 5, 661-665 (2005).

Filler, M., C. Mui, G. Wang, C. Musgrave, and S. Bent, "Competition and Selectivity in the Reaction of Nitriles on Ge (100)-2x1," *Journal of the American Chemical Society* 125, 4928-4936 (2003).

G. Serrano & N.C. Seeman, Nanotecnologia basada en ADN, Revista de Química 19, 11-20 (2004), in Spanish.

Gallego-Juárez, J.A. (1989) Piezoelectric ceramics and ultrasonic transducers. J. Phys. E: Sci. Instrum. 22, 804-816

H. Qiu, J.C. Dewan and N.C. Seeman, A DNA Decamer with a Sticky End: The Crystal Structure of d-CGACGATCGT. J. Mol. Biol. 267, 881-898 (1997).

H. Yan and N.C. Seeman, Edge-Sharing Motifs in DNA Nanotechnology. Journal of Supramolecular Chemistry 1, 229-237 (2003).

H. Yan, X. Zhang, Z. Shen and N.C. Seeman, A Robust DNA Mechanical Device Controlled by Hybridization Topology, Nature 415, 62-65 (2002).

Hoh, J.H. and Hansma, P.K. (1992) Atomic force microscopy for high-resolution imaging in cell biology. Trends Cell Bio. 2, 208-213

J. Chen and N.C. Seeman, The Synthesis from DNA of a Molecule with the Connectivity of a Cube, Nature 350, 631-633 (1991).

J. D. Le, Y. Pinto, N.C. Seeman, K. Musier-Forsyth, T. A. Taton and R.A. Kiehl, Self-Assembly of Nanoelectronic Component Arrays by In Situ Hybridization to 2D DNA Scaffolding, NanoLetters 4, 2343-2347 (2004).

J. Qi, X. Li, X.P. Yang and N.C. Seeman, The Ligation of Triangles Built from Bulged Three-Arm DNA Branched Junctions, Journal of the American Chemical Society, 118, 6121-6130 (1996).

J.H. Reif, T.H. LaBean & N.C. Seeman, Challenges and Applications for Self-Assembled DNA Nanostructures, Sixth International Workshop on DNA-Based Computers, DNA 2000, Leiden, The Netherlands, (June, 2000) ed. A. Condon, G. Rozenberg. Springer-Verlag, Berlin Heidelberg, Lecture Notes in Computer Science 2054, 173-198, (2001).

Kang, J. and C. Musgrave, "A Quantum Chemical Study of the Self-Directed Growth Mechanism of Styrene and Propylene Molecular Nanowires on the Silicon (100) 2x1 Surface," *Journal of Chemical Physics* 116, 9907-9913 (2002).

Kang, J. and C. Musgrave, "The Mechanism of Atomic Layer Deposition of SiO 2 on the Silicon (100)-2x1 Surface Using SiC1 4 and H 2 O as Precursors," *Journal of Applied Physics* 91, 3408-3414 (2002).

Keller, D.J. and Chih-Chung, C. (1992) Imaging steep, high structures by scanning force microscopy with electron beam deposited tips. Surf. Sci. 268, 333-339

L. Zhu, P.S. Lukeman, J. Canary & N.C. Seeman, Nylon/DNA: Single-Stranded DNA with Covalently Stitched Nylon Lining, J. Am. Chem. Soc. 125, 10178-10179 (2003).

M.L. Petrillo, C.J. Newton, R.P. Cunningham, R.-I. Ma, N.R. Kallenbach and N.C. Seeman. The Ligation and Flexibility of 4-Arm DNA Junctions. Biopolymers, 27, 1337-1352 (1988).

Meyer, G. and Amer, N.M. (1988) Novel optical approach to atomic force microscopy. Appl. Phys. Lett. 53(12), 1045-1047

Meyer, G. and Amer, N.M. (1990) Simultaneous measurement of lateral and normal forces with an optical-beam-deflection atomic force microscope. Appl. Phys. Lett. 57(20), 2089-2091

Mui, C. and C. Musgrave, "The Hydroxylation and Oxidation of the Ge(100)-2x1 Surface by Hydrogen Peroxide and Water," *Langmuir*, 20, 7604-7609 (2004).

Mui, C., S. Bent and C. Musgrave, "A Quantum Chemistry Based Statistical Mechanical Model of Hydrogen Deposition from Si(100)-2x1, Ge(100)-2x1, and SiGe Alloy Surfaces," *Journal of Physical Chemistry B*, 108, 18243-18253, (2004).

Mui, C., Y. Widjaja, J. Kang, and C. Musgrave, "Surface Reaction Mechanisms for Atomic Layer Deposition of Silicon Nitride," *Surface Science,* 557, 159-170, (2004).

N. Jonoska, P. Sa-Ardyen and N.C. Seeman, Computation by Self-Assembly of DNA Graphs, Journal of Genetic Programming and Evolvable Machines 4, 123-137 (2003).

N. Jonoska, S. Liao, & N.C. Seeman, Transducers with programmable input by DNA Self-assembly, Aspects of Molecular Computing, Lecture Notes in Computer Science 2340, Springer-Verlag, Berlin, 219-240 (2004).

N.C. Seeman & A.M. Belcher, Emulating Biology: Nanotechnology from the Bottom Up, Proceedings of the National Academy of Sciences (USA) 99 (supp. 2), 6451-6455 (2002).

N.C. Seeman and N.R. Kallenbach, Design of Immobile Nucleic Acid Junctions. Biophysical Journal 44, 201-209 (1983).

N.C. Seeman and N.R. Kallenbach, Nucleic Acid Junctions, The Tensors of Life? In: Nucleic Acids: The Vectors of Life, ed. by B. Pullman and J. Jortner, D. Reidel, Dordrecht, 183-200 (1983).

N.C. Seeman and N.R. Kallenbach, Nucleic Acid Junctions: A Successful Experiment in Macromolecular Design. In: Molecular Structure: Chemical Reactivity and Biological Activity, ed. by J.J. Stezowski J.-L. Huang and M.-C. Shao, Oxford University Press, Oxford, (1988) pp. 189-194.

N.C. Seeman and P.S. Lukeman, Nucleic Acid Nanostructures, Reports on Progress in Physics 68, 237-270 (2005).

N.C. Seeman, At the Crossroads of Chemistry, Biology and Materials: Structural DNA Nanotechnology, Chemistry & Biology 10, 1151-1159 (2003).

N.C. Seeman, B. Ding, S. Liao, T. Wang, W.B. Sherman, P.E. Constantinou, J. Kopatsch, C. Mao, R. Sha, F. Liu, H. Yan & P.S. Lukeman, Experiments in Structural DNA Nanotechnology: Arrays and Devices, Proc. SPIE; Nanofabrication: Technologies, Devices and Applications 5592, 71-81 (2005).

N.C. Seeman, Biochemistry and Structural DNA Nanotechnology: An Evolving Symbiotic Relationship, Biochemistry 42, 7259-7269 (2003).

N.C. Seeman, C. Mao, F. Liu, R. Sha, X. Yang, L. Wenzler, X. Li, Z. Shen, H. Yan, P. Sa-Ardyen, X. Zhang, W. Shen, J. Birac, P. Lukeman, Y. Pinto, J. Qi, B. Liu, H. Qiu, S.M. Du, H. Wang, W. Sun, Y. Wang, T.-J. Fu, Y. Zhang, J.E. Mueller and J. Chen, Nicks, Nodes, and New Motifs for DNA Nanotechnology, Frontiers of Nano-Optoelectronic Systems, ed. by L. Pavesi & E. Buzanova, Kluwer, Dordrecht, 177-198 (2000).

N.C. Seeman, De Novo Design of Sequences for Nucleic Acid Structure Engineering, Journal of Biomolecular Structure and Dynamics 8, 573-581 (1990).

N.C. Seeman, Design and Engineering of Nucleic Acid Nanoscale Assemblies, Current Opinion in Structural Biology, 6, 519-526 (1996).

N.C. Seeman, DNA in a Material World, Nature 421, 33-37 (2003).

P. Sa-Ardyen, A.V. Vologodskii and N.C. Seeman, The Flexibility of DNA Double Crossover Molecules. Biophysical Journal 84, 3829-3837 (2003).

P.S. Lukeman, A. Mittal, & N.C. Seeman, Two Dimensional PNA/DNA Arrays: Estimating the Helicity of Unusual Nucleic Acid Polymers, Chemical Communications 2004, 1694-1695 (2004).

Putman, C.A.J., De Grooth, B.G., Van Hulst, N.F., and Greve, J. (1992) A detailed analysis of the optical beam deflection technique for use in atomic force microscopy. J. App. Phys. 72(1), 6-12

R. Sha, F. Liu and N.C. Seeman, Atomic Force Measurement of the Interdomain Angle in Symmetric Holliday Junctions, Biochemistry 41, 5950-5955 (2002).

R. Sha, F. Liu, D.P. Millar N.C. Seeman, Atomic Force Microscopy of Parallel DNA Branched Junction Arrays, Chemistry & Biology 7, 743-751 (2000).

R.-I. Ma, N.R. Kallenbach, R.D. Sheardy, M.L. Petrillo and N.C. Seeman, 3-Arm Nucleic Acid Junctions Are Flexible. Nucleic Acids Research 14, 9745-9753 (1986).

Reviews in Computational Chemistry, Vol 2, Kenny B. Lipkowitz and Donald B. Boyd, VCH Publishers, 1991.

Rod Logic and Thermal Noise in the Mechanical Nanocomputer, by K. Eric Drexler, in *Proceedings of the 3rd International Symposium on Molecular Electronic Devices*, F.L. Carter, R.E. Siatkowski, H. Wohltjen, eds., North-Holland 1988.

S. Liao and N.C. Seeman, Translation of DNA Signals into Polymer Assembly Instructions, Science 306, 2072-2074 (2004).

S. Xiao, F. Liu, A. Rosen, J.F. Hainfeld, N.C. Seeman, K.M. Musier-Forsyth & R.A. Kiehl, Self-Assembly of Nanoparticle Arrays by DNA Scaffolding, J. Nanoparticle Research 4, 313-317 (2002).

Senosiain, J., C. Musgrave and D. Golden, "Temperature and Pressure Dependence of the Reaction of OH and CO: Master Equation Modeling on a High Level Potential Energy Surface," *International Journal of Chemical Kinetics* 35, 464-474 (2003).

Sha, R., Zhang, X., Liao, S., Constantinou, P.E., Ding, B., Wang, T., Garibotti, A.V., Zhong, H., Israel, L.B., Wang, X., Wu, G., Chakraborty, B., Chen, J., Zhang, Y., Mao, C., Yan, H., Kopatsch, J., Zheng, J., Lukeman, P.S., Sherman, W.B. and Seeman, N.C., Motifs and Methods in Structural DNA Nanotechnology, Proc. Intl. Conf. Nanomaterials, NANO 2005, July 13-15, 2005, Mepco Schlenk Engineering College, Srivakasi, India, V. Rajendran, ed., pp. 3-10 (2005).

T. LaBean, H. Yan, J. Kopatsch, F. Liu, E. Winfree, J.H. Reif and N.C. Seeman, The Construction of DNA Triple Crossover Molecules, Journal of the American Chemical Society 122, 1848-1860 (2000).

W. Shen, M. Bruist, S. Goodman & N.C. Seeman, A Nanomechanical Device for Measuring the Excess Binding Energy of Proteins that Distort DNA, Angew. Chem. Int. Ed. 43, 4750-4752 (2004); Angew.Chem. 116, 4854-4856 (2004).

W.B. Sherman and N.C. Seeman, A Precisely Controlled DNA Bipedal Walking Device, NanoLetters 4, 1203-1207 (2004); Erratum, 4, 1801-1801.

Weisenhorn, A.L., Hansma, P.K., Albrecht, T.R., and Quate, C.F. (1989) Forces in atomic force microscopy in air and water. Appl. Phys. Lett. 54(26), 2651-2653

Widjaja, Y. and C. Musgrave, "Atomic Layer Deposition of Hafnium Oxide: A Detailed Reaction Mechanism from First Principles," *Journal of Chemical Physics* 117, 1931-1934 (2002).

Widjaja, Y. and C. Musgrave, "Indirect Adsorbate-Adsorbate Interactions Mediated Through the Surface Electronic Structure of the Si(100)-(2x1) Surface," *Journal of Chemical Physics* 120, 1555-1559 (2004).

Widjaja, Y. and C. Musgrave, "Quantum Chemical Study of the Elementary Reactions in Zirconium Oxide Atomic Layer Deposition," *Applied Physics Letters* 81, 304-306 (2002).

X. Yang, B. Liu, A. Vologodskii, B. Kemper and N.C. Seeman, Torsional Control of Double Stranded DNA Branch Migration. Biopolymers 45, 69-83 (1998).

X. Yang, L.A. Wenzler, J. Qi, X. Li and N.C. Seeman, Ligation of DNA Triangles Containing Double Crossover Molecules, Journal of the American Chemical Society 120, 9779-9786 (1998).

X. Zhang, H. Yan, Z. Shen and N.C. Seeman, Paranemic Cohesion of Topologically-Closed DNA Molecules, J Am. Chem. Soc.124, 12940-12941 (2002).

X.J. Li, X.P. Yang, J. Qi, and N.C. Seeman, Antiparallel DNA Double Crossover Molecules as Components for Nanoconstruction, Journal of the American Chemical Society, 118, 6131-6140 (1996).

Xu, Y. and C. Musgrave, "A DFT Study of the Atomic Layer Deposition of Al 2 O 3 on SAMs: The Effect of SAM Termination," *Chemistry of Materials,* 16, 646 (2004).

Y. Wang, J.E. Mueller, B. Kemper, and N.C. Seeman, The Assembly and Characterization of 5-Arm and 6-Arm DNA Junctions, Biochemistry 30, 5667-5674 (1991).

Y. Zhang and N.C. Seeman, A Solid-Support Methodology for the Construction of Geometrical Objects from DNA, Journal of the American Chemical Society 114, 2656-2663 (1992).

Y. Zhang and N.C. Seeman, The Construction of a DNA Truncated Octahedron, Journal of the American Chemical Society 116, 1661-1669 (1994).

Glossary

At the Core: A Non-Heavy Metal Quantum Dot. Research at Evident, which is focused on advancing product development, led to the first commercially available nanocrystals that comprise a III-V semiconductor material: Indium, Gallium and Phosphide core quantum dots, with a unique molecular metallic platting compound to form the shell or InGaP/ZnS EviDots-MP™.

Bone substitutes: Synthetic or natural materials for the replacement of bones or bone tissue. They include hard tissue replacement polymers, natural coral, hydroxyapatite, beta- tricalcium phosphate, and various other biomaterials. The bone substitutes as inert materials can be incorporated into surrounding tissue or gradually replaced by original tissue.

Brownian Motion: Motion of a particle in a fluid owing to thermal agitation, observed in 1827 by Robert Brown. (Originally thought to be caused by vital force, Brownian motion in fact plays a vital role in the assembly and activity of the molecular structures of life).

Disruptive techonlogies: In his book, *The Innovator's Dilemma*, Harvard professor Clayton Christensen used the term "disruptive technologies" to describe developments capable of displacing established market leaders.

Enzymes: Molecular machines found in nature, made of protein, which can catalyze (speed up) chemical reactions.

Fail-stop: Describes a component or subsystem that, in the event of a failure, produces no output (e.g., of material or data) rather than producing a damaged or incorrect output.

Fertilization: Fusion of a male and a female gamete (both haploid) to form a diploid zygote, which develops into a new individual.

FET: Field-Effect Transistor — semiconductor device whose insulated gate electrode controls current flow.

Fiber-optic: Relates to transmission of information as modulated light in tiny transparent fibers instead of copper wires.

Fick's first law: The diffusion flux is proportional to the concentration gradient. This relationship is employed for steady-state diffusion situations. The time rate of change of concentration is proportional to the second derivative of concentration. This relationship is employed in non-steady-state diffusion situations.

Field-Effect Transistor (FET): Field-Effect Transistor — semiconductor device whose insulated gate electrode controls current flow.

Fluorescent dye: Molecule that absorbs light at one wavelength and responds by emitting light at another wavelength; the emitted light is of longer wavelength (and hence of lower energy) than the light absorbed.

Free-energy change (DG): Change in the free energy during a reaction: the free energy of the product molecules minus the free energy of the starting molecules. A large negative value of DG indicates that the reaction has a strong tendency to occur.

Grain growth: The increase in average grain size of a polycrystalline material: for most materials, an elevated temperature heat treatment is necessary.

Gray goo: The name given to free-range self-replicating miniature machines that could, in theory, run out of control and cause severe damage to the biosphere. The actual threat is generally overrated, as we explain here.

Green ceramic body: A ceramic piece, formed as a particulate aggregate, that has been dried but not fired.

Group: A set of linked atoms in a molecule; a defined substructure. Typically, a set that is usefully regarded as a unit in chemical reactions of interest.

Heme: An iron complex. Cyclic organic molecule containing an iron atom that carries oxygen in hemoglobin and carries an electron in cytochromes.

Hemidesmosome: Specialized cell junction between an epithelial cell and the underlying basal lamina.

Hemoglobin: A biomolecule composed of four myoglobine-like units (proteins plus heme) that can bind and transport four oxygen molecules in the blood.

The major protein in red blood cells that associates with O_2 in the lungs by means of a bound heme group.

Immune Machines: Medical nanomachines designed for internal use, especially in the bloodstream and digestive tract, able to identify and disable intruders such as bacteria and viruses.

Immune response: Response made by the immune system of a vertebrate when a foreign substance or microorganism enters its body.

Immune system: Population of lymphocytes and other white blood cells in the vertebrate body that defends it against infection.

Immunoglobulin (Ig): An antibody molecule. Higher vertebrates have five classes of immunoglobulin IgA, IgD, IgE, IgG, and IgM - each with a different role in the immune response.

Index of refraction (n): The ratio of the velocity of light in a vacuum to the velocity in some medium.

Inductance [in Henry, H]: That property of an electric circuit which tends to oppose change in current in the circuit. One henry (H) is the inductance of a closed circuit in which an electromotive force of 1 volt is produced when the electric current in the circuit varies uniformly at the rate of 1 ampere per second.

Induction (embryonic): Change in the developmental fate of one tissue caused by an interaction with another tissue.

Integrated Circuit (IC): An electronic circuit consisting of many interconnected devices on one piece of semiconductor, typically into 10 millimeters on a side. ICs are the major building blocks of today's computers. Semiconductor circuit, typically on a very small silicon chip, containing microfabricated transistors, diodes, resistors, capacitors, etc.

Intelligent Agent: Aka "software agent". Software that can do things without supervision, because it knows your patterns, history, preferences, likes, dislikes, and so forth. You want to take a vacation - it knows that you really enjoyed that trip to Hawaii, and that you prefer to fly at night, 1st class. It also knows that the bungalow you rented last time was marked as being 5-star, and worth a re-visit. Your IA then collates all your parameters, searches the internet for flights, car rentals, restaurant reservations, and lodgings, and schedules everything for you, with options on the side. No more travel agent - you have a software agent to handle things! Many experts agree that by 2010 we will each have one, and that they will greatly reduce our daily load of trivial and redundant tasks.

Interagency Working Group on Nanoscience, Engineering and Technology IWGN, National Science and Technology Council NSTC, US

Interface: In physical terms, an interface is the boundry between two phases, for instance between a solid and liquid or between a liquid and gas.

Intermolecular: Between more than one molecule. Describes an interaction (e.g., a chemical reaction) between different molecules.

Internal energy: The sum of the kinetic and potential energies (including electromagnetic field energies) of the particles that make up a system.

Internet: Worldwide digital communication network in which packets of information travel between senders and recipients.

Interstitial diffusion: a diffusion mechanism that causes atomic motion from interstitial site to interstitial site.

Intramolecular: Describes an interaction (e.g., a chemical reaction) within a single molecule. Intramolecular interactions between widely separated parts of a molecule resemble intermolecular interactions in most respects. (2) Within a single molecule.

Intrinsic: Characterizes pure undoped semiconductor; electrical conductivity depends only on temperature and the band gap energy.

Ion: An atom or molecule with a net charge. **(2)** An atom with more or fewer electrons than those needed to cancel the electronic charge of the nucleus. An ion is an atom with a net electric charge.

Ionic bond: A chemical bond resulting chiefly from the electrostatic attraction between positive and negative ions. A coulombic interatomic bond existing between two adjacent and oppositely charged ions.

Kilobyte (kB): 2^{10} (= 1024, or about one thousand) bytes of information.

Kilocalorie (kcal): Unit of heat energy equal to 1000 calories. Often used to express the energy content of food or molecules: bond strengths, for example, are measured in kcal/mole. An alternative unit in wide use is the kilojoule, equal to 0.24 kcal.

Kilohertz (kHz): One thousand cycles per second.

Kilojoule: Standard unit of energy equal to 1000 joules, or 0.24 kilocalories.

Kinesin: One type of motor protein that uses the energy of ATP hydrolysis to move along a microtubule.

Kinetic energy: Energy resulting from the motion of masses.

Lamins: Intermediate filament proteins that form the fibrous matrix (nuclear lamina) on the inner surface of the nuclear envelope.

Langmuir-Blodgett (LB) Film: A film of organic material (often surfactant molecules) assembled at the liquid-gas interface and subsequently transferred onto a solid substrate.

Langmuir-Blodgett: The name of a nanofabrication technique used to create ultrathin films (monolayers and isolated molecular layers), the end result of which is called a "Langmuir-Blodgett film".

Laser trimming: A method for adjusting the value of thin- or thick-film resistors by using a computer-controlled laser system.

Laser tweezers: Laser tweezers and related approaches are important for manipulation and isolation of subcellular *organelles* and structural components. [National Center for Research Resources "Integrated Genomics Technologies Workshop Report" Jan 1999]

Laser: Light Amplification by the Stimulated Emission of Radiation — quantum device that produces coherent light.

LCD (Liquid Crystal Display): LCD is the predominant technology used in flat panel displays. The principle that makes the display work is this: A crystalís alignment can be altered with an electric current. If the crystal is lined up one way ñ it will allow the light waves to pass through a polarized filter, but if the electric current alters the crystalís alignment, it will guide light so that the polarized filter blocks the light. By densely packing red, blue and green light emitting crystals next to each other on a sheet (ìcalled a substrateî), one can create a full color display. The great thing about LCD is that the crystals can be packed together closely, allowing for a higher-resolution, finer-detail display. The con is that LCDs are somewhat fragile, require a lot of power and are relatively less bright. Liquid Crystal Display — display device employing light source and electrically alterable optically active thin film.

Leading strand: One of the two newly made strands of DNA found at a replication fork. The leading strand is made by continuous synthesis in the 5′-to-3′ direction.

Leakage: The loss of all or parts of a useful agent, as of the electric current that flows through an insulator or the magnetic flux that passes outside useful flux circuits.

Lectin: Protein that binds tightly to a specific sugar. Abundant lectins derived from plant seeds are often used as affinity reagents to purify glycoproteins or to detect them on the surface of cells.

LED: Light-Emitting diode — semiconducting diode that produces visible or infrared radiation.

Macroscale: Larger than nanoscale; often implies a design that humans can directly interact with; too large to be built by a single assembler (one cubic micron of diamond contains 176 billion atoms).

Macrosensing: In medical nanorobotics, the detection of global somatic states (inside the human body) and extrasomatic states (sensory data originating outside of the human body) by in vivo nanorobots.

Magnetic field strength (designated by H) [A/m]: Magnetic field produced by a current, independent of the presence of magnetic material. The units of H are ampere-turns per meter, or just amperes per meter.

Magnetic flux density or magnetic induction (designated by B): The magnetic field produced in a substance by an external magnetic field. The units of B are tesla (T). One tesla is the magnetic flux density given by a magnetic flux of 1 weber per square meter. One weber is a magnetic flux that, linking a circuit of 1 turn, would produce in it an electromotive force of 1 volt if it were reduced to zero at a uniform rate in 1 second. Both B and H are field vectors. One henry (H) is the inductance of a closed circuit in which an electromotive force of 1 volt is produced when the electric current in the circuit varies uniformly at the rate of 1 ampere per second. The magnetic field strength and flux density are related according to: $B = \mu H$, where μ is the permeability (see under permeability).

Magnetic Force Microscopy (MFM): A method for observing local magnetic fields near a surface by scanning the surface with a magnetic probe.

Magnetic susceptibility ($\div_m$): The proportionality constant between the magnetization M (see under"magnetization") and the magnetic field strength H. The magnetic susceptibility is unitless.

Magnetization (M): The total magnetic moment per unit volume of material. Also, a measure of the contribution to the magnetic flux by some material within an H field. The magnitude of M is proportional to the applied field as: $M = \div_m \times H$, with $\div_m$ the magnetic susceptibility.

Magnetostrictive material: A material that changes dimension in the presence of a magnetic field or generates a magnetic field when mechanically deformed.

Manufacturing: The ability to make products, in this case ranging from clothing, to electronics, to medical devices, to books, to building materials, and much more.

Martensite: A metastable iron phase supersaturated in carbon that is the product of a diffusionless (athermal) transformation from austenite.

Mask: Pattern on glass, like a photographic negative, for producing integrated-circuit elements on semiconductor wafer.

Mass to Charge Ratio: A number defining how a particle will respond to an electric or magnetic field that can be calculated by dividing the mass of a particle by its charge.

Massometer: In medical nanorobotics, a nanosensor device for measuring the mass of individual molecules or small physical objects to single-proton resolution.

Materials science: Science of ceramics, glass, metals, plastics, semiconductors.

MatML Materials Markup Language: Materials property data distributed on the World Wide Web in documents using hypertext markup language.

Matter as Software: "Autonomous, motile microdevices clearly are on the horizon. They may be regarded as the first step in the evolution of a technology for "programming" the structure and properties of material objects at the microscopic and the submicroscopic levels. As this evolution progresses, *the physical and economic properties of such programmable matter are likely to become much like those of present day software."* [MITRE Corporation]

MCM: MultiChip Module; the interconnection of two or more semiconductor chips in a semiconductor-type package.

Measurand: A physical quantity, condition, or property that is to be measured.

Meat Machine: AKA Cabinet Beast. A box containing assemblers and raw material, within which is formed meat [or whatever else it was programmed to make].

MEMS: Stood originally for Micro-ElectroMechanical System — microscopic mechanical elements, fabricated on silicon chips by techniques similar to those used in integrated circuit manufacture, for use as sensors, actuators, and other devices. Today almost any miniaturized device (based on Si technology or traditional precision engineering, chemical or mechanical) is referred to as a MEMS device.

Mesoscale: A device or structure larger than the nanoscale (10^-9 m) and smaller than the megascale; the exact size depends heavily on the context and usually ranges between very large nanodevices (10^-7 m) and the human scale (1 m). [AS]

Messenger molecule: A chemically recognizable molecule which can convey information after it is received and decoded by an appropriate chemical sensor.

Metal nanoshells: A new type of *nanoparticle* composed of a *semiconductor* or dielectric core coated with an ultrathin conductive layer.. By adjusting the relative core and shell thicknesses, metal nanoshells can be fabricated that will absorb or scatter light at any wavelength across the entire visible and infrared range of the electromagnetic spectrum.

Metamorphic: In medical nanorobotics, capable of adopting multiple physical configurations via smooth changes from one configuration to another.

Metastable: A classical system is metastable if it is above its minimum-energy state, but requires an energy input before it can reach a lower-energy state; accordingly, a metastable system can act like a stable system, provided that energy inputs (e.g., thermal fluctuations) remain below some threshold. Systems with strong metastability are commonly described as stable. Quantum mechanical effects can permit metastable states to reach lower energies by tunneling, without an energy input; an associated, broader definition of *metastable* embraces all systems that have a long lifetime (by some standard) in a state above the minimum-energy state.

Micellar nanocontainers: Block copolymer micelles are water- soluble biocompatible nanocontainers with great potential for delivering hydrophobic drugs. An understanding of their cellular distribution is essential to achieving selective delivery of drugs at the subcellular level. R. Savic, L. Luo, A. Eisenberg, D. Maysinger, Micellar nanocontainers distribute to defined cytoplasmic organelles Science 300 (5619): 615- 618, Apr. 25, 2003

Micelles: Micelles are small, spherical structures composed of molecules that attract one another to reduce surface tension. The head of the molecule is hydrophilic, meaning it likes water, while the interior portion is hydrophobic, meaning it avoids water.

Microbiotagraphics: Mapping the microbiotic populations present in the human body.

Microbubbles: Very small encapsulated gas bubbles (diameters of micrometers) that can be used in diagnostic and therapeutic applications. Upon exposure to sufficiently intense ultrasound, microbubbles will cavitate, rupture, disappear, release gas content, etc. Such characteristics of the microbubbles can be used to enhance diagnostic tests, dissolve blood clots, and deliver drugs or genes for therapy. MeSH 2004

Microchemistry: The development and use of techniques and equipment to study or perform chemical reactions, with small quantities of materials, frequently less than a milligram or a milliliter. MeSH 2003

Microdevices: Narrower terms: integrated microdevices, nanodevices, SED Single Electron Devices. Related terms microelectronics, microfluidics, micro- TAS.

Micro-Electromechanical Systems (MEMS): "The fabrication or micro-machining of materials to make stationary and moving structures, devices and systems of a nominal size scale from a few centimeters to a few micrometers."

Microelectronics: Narrower terms: MEMS, nanoelectronics, optoelectronics, SED Single Electron Devices. Related terms: molecular electronics, semiconductors

Microencapsulation: Individually encapsulated small particles.

Microinjection: The insertion of a substance into a cell through a microelectrode. Typical applications include the injection of drugs, histochemical markers (such as horseradish peroxidase or lucifer yellow) and RNA or DNA in molecular biological studies. To extrude the substances through the very fine electrode tips, either hydrostatic pressure (pressure injection) or electric currents (ionophoresis) is employed. [OMD] A technique fo: introducing a solution of DNA, protein, or other soluble material into a cell using a fine microcapillary pipet. [Life Sciences Dictionary]

MNT: An abbreviation for molecular nanotechnology; refers to the concept of building complicated machines out of precisely designed molecules.

Mobility (electron, and hole): The proportionality constant between the carrier drift velocity and applied electric field.

Modulus of elasticity (E) : The ratio of stress to strain when deformation is totally elastic. Also the Young's modulus.

MOEMS MicroOpticalElectroMechanical systems: In addition to mechanical and electrical components, integrate waveguides or other optical features into the body of the silicon chip. Intellisense Corp. MEMS The Next Small Thing Google = about 10 Aug. 8, 2002; about 12 June 23, 2004; about 10 May 2, 2005

Moiety: A portion of a molecular structure having some property of interest.

Molality: The molality or molal concentration (symbol m) is the amount of substance per unit mass of solvent or mol kg^{-1}. Concentration in a liquid solution (symbol c), in terms of the number of moles of a solute dissolved in 10^6 mm^3 (10^3 cm^3) of solution in mol l^{-1}.

Molding (plastics): Shaping a plastic material by forcing it, under pressure at a high temperature, into a mold cavity.

Mole: A number of instances of something (typically a molecular species) equaling ~6.022 1023. *Mole* ordinarily means gram-mole; a kilogram-mole is ~6.022 1026.

Molecular - Nanotechnology: The technology of precisely- constructed molecular-scale machines; from nanometer: a billionth of a meter. For a good introduction, see What is Nanotechnology? or Ralph C. Merkle's nanotechnology page.

Molecular assembler: A general-purpose device for molecular manufacturing, able to guide chemical reactions by positioning individual molecules to atomic accuracy (e.g. mechanosynthesis) and to construct a wide range of useful and stable molecular structures according to precise specifications. Also known as an assembler, a molecular assembler is a molecular machine that can build a molecular structure from its component building blocks. [ZY]

Molecular Beam Epitaxy: [MBE] Process used to make compound (multi-layer) semiconductors. Consists of depositing alternating layers of materials, layer by layer, one type after another (such as the semiconductors gallium arsenide and aluminum gallium arsenide).

Molecular Integrated Microsystems (MIMS): microsystems in which functions found in biological and nanoscale systems are combined with manufacturable materials. See Molecular Integrated Microsystems

Molecular self-assembly: The spontaneous formation of molecules into covalently bonded, well-defined, stable structures - is a very important concept in biological systems and has increasingly become a focus of non- biological research.

Molecular sorting rotor: A class of nanomechanical device capable of selectively binding (or releasing) molecules from (or to) solution, and of transporting these bound molecules against significant concentration gradients.

Molecular surgery (molecular repair): In medical nanorobotics, the analysis and physical correction of molecular structures in the body using medical nanomachines. Analysis and physical correction of molecular structures in the body using medical nanomachines.

Molecular Systems Engineering: Design, analysis, and construction of systems of molecular parts working together to carry out a useful purpose.

Molecular Wire: A molecular wire - the simplest electronic component - is a quasi-one-dimensional molecule that can transport charge carriers (electrons or holes) between its ends. [Michael D Ward]

Molecularly imprinted polymers MIPs: A new class of materials that have artificially created receptor structures. Since their discovery in 1972, MIPs have attracted considerable interest from scientists and 3engineers involved with the development of chromatographic absorbents, membranes, sensors and enzyme and receptor mimics.

Moledulcar manufacturing: The automated building of products from the bottom up, molecule by molecule, with atomic precision. This will make products that are extremely lightweight, flexible, durable, and potentially very 'smart'.

Monkeywrenching: In medical nanorobotics, the mechanical or chemical jamming of cellular equilibrium processes, with a cytocidal objective.

Monoclonal antibodies: Produced by injecting animals to elicit a response from lymphocytes to produce antibodies. Lymphocytes which produce antibodies with strong binding capability can be isolated and used to produce only one kind of antibody (monoclonal) on a permanent basis once the lymphocytes are immortalized. This is accomplished by fusing them (combining them genetically) with cancer cells which have the distinction of living indefinitely in a culture. Monoclonal antibodies can be produced repeatedly and collected for use in immunodetection.

Mutation: An inheritable modification in a genetic molecule, such as DNA. Mutations may be good, bad, or neutral in their effects on an organism; competition weeds out the bad, leaving the good and the neutral.

MUX: Device for combining several signals or data streams into a single flow.

Myosin: Myosin is one of the two major proteins responsible for contraction of muscle. In muscle cells, myosin is arranged in long filaments that lie parallel to the microfilaments of actin. In muscle contraction, filaments of actin and myosin alternately unlink and chemically link in a sliding action. The energy for this reaction is supplied by adenosine triphosphate.

NAD+ (nicotine adenine dinucleotide): Coenzyme that participates in an oxidation reaction by accepting a hydride ion (H-) from a donor molecule. The NADH formed is an important carrier of electrons for oxidative phosphorylation. Coenzyme closely related to NAD+ that is used extensively in biosynthetic, rather than catabolic, pathways.

Nanapheresis: In medical nanorobotics, the removal of bloodborne medical nanorobots from the body using aphersis-like processes.

Nanocentrifuge: In medical nanorobotics, a proposed nanodevice that can spin materials at very high speed, imparting rotational accelerations of up to one trillion gravities (g's), thus permitting rapid sortation.

Nanochip: In Friday's issue [Aug. 1, 2001] of the journal *Science,* physicists from IBM's Thomas J. Watson Research Center [Philip G. Collins, Michael S. Arnold and Phaedon Avouris] announce their fabrication of the world's first array of transistors made from carbon nanotube. We are approaching the limits of standard microchip technology; thus, the "nanochip" — a next-smaller microchip. [ed] They are also a next-gen device for mass storage, of significantly higher density, with greater speed, and much lower cost.

Nanochondria: Nanomachines existing inside living cells, participating in their biochemistry (like mitochondria) and/or assembling various structures.

Nanocontainers: "Micellar nanocontainers" or "Micelles," these are nanoscale polymeric containers that could be used to selectively deliver hydrophobic drugs to specific sites within individual cells.

Nanocrit (Nct)**:** In medical nanorobotics, volume-fraction or bloodstream concentration of medical nanorobots, expressed as a percentage.

Nanomachining: like traditional machining, where portions of the structure are removed or modified, nanomachining involves changing the structure of nano-scale materials or molecules.

Nanomanipulation: The process of manipulating items at an atomic or molecular scale in order to produce precise structures.

NanoManipulator: uses virtual reality (VR) goggles and a force feedback probe as an interface to a scanning probe microscope, providing researchers with a new way to interact with the atomic world. Researchers can travel over genes, tickle viruses, push bacteria around, and tap on molecules - the nanoManipulator simplifies the process and allows researchers to play with their atoms. University of North Carolina at Chapel Hill (UNC-CH) *The Nanomanipulator* from the Center for Computer Integrated Systems for Microscopy and Manipulation (CISMM) at UNC Chapel Hill. Part of the Nanoscale Science Research Group (NSRG).

Nanopens & Nanopencils: (AKA: Atomic Pencil) *"Analogous to using a quill pen but on a billionth the scale"*, and may transform dip-pen nanolithography. Allows for drawing electronic circuits a thousand times smaller than current ones. The "pen" is an atomic force microscope (AFM).

NanoPGM-nanometer-scale patterned granular motion: The goal of NanoPGM is to generate millions of ìnanofingers, finger-like structures each only a few nanometers long, that might someday perform precise, massively parallel manipulation of molecules and directed assembly of other nanometer-scale objects. This ability answers one of the biggest technical challenges facing builders of nanocomputers: how to arrange as many as a trillion molecular computing components in an area only a few millimeters square. [MITRE / Alex Wissner-Gross]

Nanopharmaceuticals: Nanoscale particles used to modulate drug transport for drug uptake and delivery applications.

Nanophase Carbon Materials (carbon nanotubes, nanodiamond, nanocomposite]: A form of matter in which small clusters of atoms form the building blocks of a larger structure. These structures differ from those of naturally occurring crystals, in which individual atoms arrange themselves into a lattice.

Nanosensor: A chemical or physical sensor constructed using nanoscale components, usually microscopic or submicroscopic in size.

Nanoshells: Nanoscale metal spheres, which can absorb or scatter light at virtually any wavelength.

"The nanoshells act as an amazingly versatile optical component on the nanometer scale: they may provide a whole new approach to optical materials and components," Professor Naomi Halas.

Nanosieving: In medical nanorobotics, a nanodevice that can sort molecules or other nanoscale objects by physical sieving.

Nanosieving: In medical nanorobotics, a nanodevice that can sort molecules or other nanoscale objects by physical sieving.

Nanosome: Nanodevices existing symbiotically inside biological cells, doing mechanosynthesis and disassembly for it and replicating with the cell. Similar to nanochondria. [AS January 1998]

Nanosome: Nanodevices existing symbiotically inside biological cells, doing mechanosynthesis and disassembly for it and replicating with the cell. Similar to nanochondria. [Anders Sandberg January 1998]

Nanosources: sources that emit light from nanometre-scale volumes.

Nanospheres: Self-assembling nanospheres that fit inside each other like Russian dolls are one form of a broad range of submicroscopic spheres ... The durable silica spheres, which range in size from 2 to 50 nanometers, form in a few seconds, are small enough to be introduced into the body, and have uniform pores that could enable controlled release of drugs. The spheres can absorb organic and inorganic substances including small particles of iron, which means they can be controlled by magnets and the contents released as needed. [Sandia National Labs news release "Self- assembled spheres may be helpful against disease or terrorism".

Nanosprings: A nanowire wrapped into a helix. Speculation is that they "may someday make highly sensitive magnetic field detectors, perhaps finding application in hard drive read heads. Alternatively, nanosprings could serve as positioners, or even as tiny conventional springs, for nanomachines of the future."

Nanosurgery: A generic term including molecular repair and cell surgery.

Nanoswarm: UFog and Goo

Nanosystem: A eutactic set of nanoscale components working together to serve a set of purposes; complex nanosystems can be of macroscopic size.

Nanotech: Slang for nanotechnology.

Nanotechism: the religion of nanotech, as opposed to the science of nanotech

Nanoterrorism: using MNT derived nanites to do damage to people or places.

Nernst equation: An equation relating the potential of an electrochemical cell to the concentrations of the cell components: $E = Eo + RT/zF \ln C_1/C_2$ with z the charge exchanged at the electrode and C_1 and C_2 concentrations of two electro-active compounds.

Neroprosthesis: Implanted cybernetic brain augmentation.

Neural Simulation: Imitating the functions of a neural system—such as the brain—by simulating the function of each cell.

Neurocomputation: The study of how natural and artificial neural networks process information.

Neuron Star: A neutron star used as a basis for a densely packed mind (similar to a jupiter brain or a omegon) exploiting neutronium or quark matter for computation [Damien Broderick, 1996]

Neuron: A nerve cell, such as those found in the brain.

Neuronaut: A person who explores their own neural functioning and internal mentational processes by various means, including deep introspection and meditation, psychoactive drugs, mind machines, and neuroscientific understanding.

Neurosuspension: Cryonically suspending only the head or brain of a person.

NMOS: An acronym for n-*channel metal-oxide-semiconductor*, as in NMOS transistor and NMOS logic.

Nono: A prefix meaning ten to the minus ninth power, or one billionth.

Nootropic: A cognition-enhancing drug that has no significant side-effects.

NOR: NOT-OR — logic gate whose output is the negation of that of the OR gate.

NOT: Logic gate whose output is binary 1 when its input is 0, and whose output is a 0 when its input is a 1.

Now shock: Being shocked or confused by the rapid changes that has already taken place, a kind of future shock before the present. [Michael Rothschild 1995, "Cornucopia Or Black Hole?", *Upside*]

NRAM™ - Nanotube-based/Nonvolatile RAM, developed by Nantero, using proprietary concepts and methods derived from leading-edge research in nanotechnology.

n-type: Characterizes a semiconductor containing predominantly mobile electrons (also see"p-type").

Pathogen (adjective pathogenic): An organism or other agent that causes diseases.

PCR (polymerase chain reaction): Technique for amplifying specific regions of DNA by multiple cycles of DNA polymerization, each followed by a brief heat treatment to separate complementary strands.

Peptide bond: Chemical bond between the carbonyl group of one amino acid and the amino group of a second amino acid - a special form of amide linkage.

Peptide map: Characteristic two-dimensional pattern (on paper or gel) formed by the separation of the mixture of peptides produced by the partial digestion of a protein.

Phagocytosis: Process by which particulate material is endocytosed ("eaten") by a cell. Prominent in carnivorous cells, such as *Amoeba proteus*, and in vertebrate macrophages and neutrophils. (From Greek *phagein*, to eat.)

Pharmacyte: In medical nanorobotics, a theorized (nanorobotic) device capable of delivering precise doses of biologically active chemicals to individually-addressed human body tissue cells (e.g. cell-by-cell drug delivery).

Phase shift: A time difference between the input and output signals.

Piezoelectric material: A ferroelectric material in which an electrical potential difference is created due to mechanical deformation, or conversely, in which the application of a voltage causes dimensional changes in the material.

Piezoelectric: A piezoelectric material expands or contracts when a voltage is applied across it. Piezoelectric tubes are often used to move the tip in an AFM or STM. If a piezoelectric crystal is compressed, a voltage appears across it. This effect is sometimes used to measure displacements. A commonly used piezoelectric is PZT $(Pb,Zr)TiO_3$.

Pinhole: The term pinhole embraces a wide variety of oxide defects and is used in a broad sense today. Listed in this category are cracks caused by thermal contraction after oxidation or by handling, and regions of oxide with low dielectric strength caused by dust particles, inadequate masking, contamination, or poor resist adhesion.

Pin-out: Diagram showing for electronic components the relations between connecting pins and internal components.

Pitting: A form of very localized corrosion wherein small pits or holes form, usually in a vertical direction.

Pixel: Picture Element — smallest element of an image, such as a dot on a computer monitor screen.

pK value: A measure of the strength of an acid on a logarithmic scale. The pK value is given by log10 (1/Ka), where Ka is the acid dissociation constant pK values often are used to compare the strengths of different acids.

Plastic deformation: Permanent or nonrecoverable deformation, accompanied by permanent atomic displacement.

Plasticizer: A low molecular weight polymer additive that enhances flexibility and workability and reduces stiffness and brittleness.

Point defect: A crystalline defect associated with one or several atomic sites.

Poisson's ratio (í): For elastic deformation, the negative ratio of lateral and axial strains that result from an applied axial stress.

Poisson's ratio: A bar of an isotropic, elastic material ordinarily shrinks laterally when it is stretched longitudinally. The lateral contracting strain divided by the applied tensile strain is Poisson's ratio, which varies from material to material.

Polarization (P): The total electric dipole moment per unit volume of dielectric material.

Polyclonal antibodies: Antibodies produced by an animal's white blood cells (lymphocytes, specifically) in response to an antigen. This response occurs naturally or can purposely be created by injecting an animal, such as a rabbit or goat, with a specific antigen. More than one kind of anti-body is produced since more than one lymphocyte is producing antibodies. This is referred to as"polyclonal". The polyclonal antibodies are isolated from the animal and can be used for detection purposes. Because the antibodies are actually a mixture with different affinities (binding capability) for the antigen of interest, some variability in performance can occur from one test to another or one batch of antibodies to another.

Polycyclic: A cyclic structure contains rings of bonds; a structure having many such rings is termed polycyclic. In the polycyclic structures of interest in this volume, a large fraction of the atoms are members of multiple small rings, resulting in considerable rigidity.

Polymer: A molecule consisting of many smaller sub-units covalently linked together.

Polymerase Chain Reaction (PCR): A chemical reaction that uses the polymerase enzyme to carry out *in vitro* replication of DNA.

Potentiometric Device: Monitors the voltage between a sensing electrode and a reference electrode. A high input impedance voltmeter is used to minimize current flow. The voltage typically is proportional to the logarithm of the analyte concentration.

Power [W]: Product of voltage and current in a component; also, refers to the field of electric energy supply.

Precision: The degree of reproducibility among several independent measurements of the same true value under specified conditions.

Presentation semaphore: In medical nanorobotics, a mechanical device used to display specific antigens, chemical ligands, or other molecular objects to the external environment, with the purpose of selectively modifying the chemical or other surface characteristics of a nanorobot exterior.

Printed circuit board: PCB — selectively metallized insulating sheet for supporting and interconnecting circuit components.

Probability density function: Consider an uncertain physical property and a corresponding space describing the range of values that the property can have (e.g., the configuration of a thermally excited N particle system and the corresponding $3N$ dimensional configuration space). The probability density function associated with a property is defined over the corresponding space; its value at a particular point is the probability per unit volume that the property has a value in an infinitesimal region around that point.

Protein Design, Protein Engineering: The design and construction of new proteins; an enabling technology for nanotechnology.

Protein Folding: "The process by which proteins acquire their functional, preordained, three-dimensional structure after they emerge, as linear polymers of amino acids, from the ribosome." [The Scientist]

Protein: Living cells contain many molecules that consist of amino acid polymers folded to form more-or-less definite three-dimensional structures; these are termed proteins. Short polymers lacking definite three-dimensional structures are termed peptides. Many proteins incorporate structures other than amino acids, either as covalently attached side chains or as bound ligands. Molecular objects made of protein form much of the molecular machinery of living cells.

Proteomics: The study of the full expression of proteins in an organism The term proteome refers to all the proteins expressed by a genome, and thus proteomics involves the identification of proteins in the body and the determination of their role in physiological and pathophysiological functions. ... Ultimately it is believed that through proteomics new disease markers and drug targets can be identified that will help design products to prevent, diagnose and treat disease. [e-proteomics.net]

Quantum Cryptography: A system based on quantum- mechanical principles. Eavesdroppers alter the quantum state of the system and so are detected. Developed by Brassard and Bennett, only small laboratory demonstrations have been made.

Quaternary Structure: Three-dimensional relationship of the different polypeptide chains in a protein complex.

Qubit: The quantum computing analog to a bit. Qubits exhibit superposition. Thus, unlike normal bits, qubits can be both 1 and 0 at the same time.

Resistor: Energy dissipative element consisting of a poor conductor in series with connecting wires.

Resolution: The minimum distance between two objects that can be distinguished in microscopy or the minimum spacing between two features that can be fabricated with lithography. The smallest measurable change in input that will produce a small but noticeable change in the output. In the context of chemical separations, defines the completeness of separation.

Resonant frequency: The frequency at which a moving member or a circuit has a maximum output for a given input.

Reverse bias: The insulating bias for a p-n junction rectifier; electrons flow into the p side of the junction.

RF: Radio Frequency — refers to alternating voltages and currents having frequencies between 9 kHz and 3 MHz.

Ribonuclease: An enzyme that cuts RNA molecules into smaller pieces.

Ribosome: A molecular machine, found in all cells, which builds protein molecules according to instructions read from RNA molecules. Ribosomes are complex structures built of protein and RNA molecules. A naturally occurring molecular machine that manufactures proteins according to instructions derived from the cell's genes.

Rigid structure: As used in this volume, a covalent structure that is reasonably stiff. In a typical rigid structure, all modes of deformation encounter first-order restoring forces resulting from some combination of bond stretching and angle bending; such a structure cannot undergo deformation by bond torsion alone. Meeting this condition usually requires a polycyclic diamondoid structure.

RNA: Ribonucleic acid; a molecule similar to DNA. In cells, the information in DNA is transcribed to RNA, which in turn is "read" to direct protein construction. Some viruses use RNA as their genetic material.

Step response: The response of a system to an instantaneous jump in the input signal.

TATA box: Consensus sequence in the promoter region of many eucaryotic genes that binds a general transcription factor and hence specifies the position where transcription is initiated.

Technocyte: A nanoscale artificial device (especially a nanite) in the human bloodstream used for repairs, cancer protection, as an artificial immune system or for other uses. [AS 1995]

Tight junction: Cell-cell junction that seals adjacent epithelial cells together, preventing the passage of most dissolved molecules from one side of the epithelial sheet to the other.

Tight-receptor structures: A receptor structure in which a bound ligand of a particular kind is confined on all sides by repulsive interactions (note that favorable binding energies are compatible with repulsive forces). A tight-receptor structure discriminates strongly against all molecules larger than the target.

Time constant: The time it takes for the output change to reach 63% of its final value.

Triple bond: A double bond is formed when a pi bond is superimposed on a single bond; adding a second pi bond results in a triple bond. The two pi bonds have perpendicular nodal planes, and their sum has roughly cylindrical symmetry, permitting rotation in much the same manner as a single bond.

Virtual reality system: A combination of computer and interface devices (goggles, gloves, etc.) that presents a user with the illusion of being in a three dimensional world of computer-generated objects.

Viscoelasticity: A type of deformation exhibiting the mechanical characteristics of viscous flow and elastic deformation.

Viscosity: The ratio of the magnitude of an applied shear stress to the velocity gradient that it produces; in other words: a measure of a noncrystalline material's resistance to permanent deformation.

Volitional normative model of disease: In medical nanorobotics, disease is said to be present in a human being upon either (1) the failure of optimal physical (e.g. biological) functioning, or (2) the failure of desired (by the patient) functioning.

Voltage [V]: Potential difference between two points: energy to move a 1-C charge through a 1-V potential difference is 1-J.

Wave function: In quantum mechanics, a complex function extending over the configuration space of a system; its complex conjugate yields the probability density function, and other mathematical operations yield other physical quantities.

Weber: Unit of magnetic flux. One weber is a magnetic flux that, linking a circuit of 1 turn, would produce in it an electromotive force of 1 volt if it were reduced to zero at a uniform rate in 1 second.

Weight percent (wt%): Concentration specification on the basis of weight (or mass) of a particular element relative to the total alloy weight (or mass).

Western blotting: Technique by which proteins are separated and immobilized on a paper sheet and then analyzed, usually by means of a labeled antibody.

Bibliography

F. Buot, "Mesoscopic Physics and Nanoelectronics: Nanoscience and Nanotechnology," *Physics Reports*, pp.73-174, 1993.

Feynman, R., "There's Plenty of Room at the Bottom: An invitation to Enter a New Field of Physics," Talk at the Annual Meeting of the American Physical Society, 29 December 1959. Reprinted in Appendix B of Crandall and Lewis. See citation above.

Fielder, F.A.; Reynolds, G.H.: 1994, 'Legal Problems of Nanotechnology: An Overview', *Southern California Interdisciplinary Law Journal*, 3, 593-629.

Fleischer, Th.; Decker, M. & Fiedeler, U. (eds.): 2004, *Große Aufmerksamkeit für kleine Welten ' Nanotechnologie und ihre Folgen*, special issue of *Technikfolgenabschätzung ' Theorie und Praxis*, 13 (2), 5-85 [www.itas.fzk.de/tatup/042/inhalt.htm].

Fogelberg, H.: 2003, 'The Material Culture of Nanotechnology', in: H. Fogelberg & H. Glimell, *Bringing Visibility to the Invisible: Towards A Social Understanding of Nanotechnology*, Göteborg: Göteborg University, pp. 99-114.

Frazier, G., "An Ideology For Nanoelectronics," in *Concurrent Computations: Algorithms, Architecture, and Technology*, Plenum Press, New York, 1988.

Glimell, H.: 2001, 'Challenging Limits ' Excerpts from an Emerging Ethnography of Nano Physicists', in: H. Glimell & O. Johlin (eds.), *The Social Production of Technology: On the Everyday Life with Things*, Göteburg, SE: BAS Publisher, chapter 7, pp. 111-131 (reprinted in: H. Fogelberg & H. Glimell, *Bringing Visibility to the Invisible: Towards A Social Understanding of Nanotechnology*, Göteborg: Göteborg University, pp. 115-139.

Glimell, H.: 2001, 'Dynamics of the Emerging Field of Nanoscience', in: M.C. Roco and W.S. Bainbridge (eds.), *Societal Implications of Nanoscience and Nanotechnology*, Dordrecht: Kluwer, pp. 156-160.

Glimell, H.: 2003, 'A Nano Narrative: The Mircopolitics of a New Generic Enabling Technology', in: H. Fogelberg & H. Glimell, *Bringing Visibility to the Invisible: Towards A Social Understanding of Nanotechnology*, Göteborg: Göteborg University, pp. 55-77.

Glimell, H.: 2003, 'Dynamics of the Emerging Field of Nanoscience', in: H. Fogelberg & H. Glimell, *Bringing Visibility to the Invisible: Towards A Social Understanding of Nanotechnology*, Göteborg: Göteborg University, pp. 79-85 [www.sts.gu.se/publications/STS_report_6.pdf].

Glimell, H.: 2004, 'Grand Visions and Lilliput Politics: Staging the Exploration of the 'Endless Frontier", in: D. Baird, A. Nordmann & J. Schummer (eds.), *Discovering the Nanoscale*, Amsterdam: IOS Press, pp. 231-246.

Goldhaber-Gordon, D., Montemerlo, M.S., Love, J.C., Opiteck, G.J., and Ellenbogen, J.C., "Overview of Nanoelectronic Devices," submitted to the *Proceedings of the IEEE*, February 1997. For more information, please send e-mail to nanotech@mitre.org.

Jounet, C., W.K. Maser, P. Bernier, A. Loiseau, M. Lamy de la Chapelle, S. Lefrant, P. Deniard, R. Lee, and J.E. Fischer. 1997. *Nature* 388:756.

K.E. Drexler, *Nanosystems: Molecular Machinery, Manufacturing, and Computation,* Wiley, New York, 1992.

Karger, J., and D.M. Ruthven. 1992. *Diffusion in zeolites*. New York: J. Wiley.

Keiper, A.: 2003, 'The Nanotechnology Revolution', *The New Atlantis*, 2, 17-34.

Khushf, G.: 2004, 'A Hierarchical Architecture for Nano-scale Science and Technology: Taking Stock of the Claims About Science Made By Advocates of NBIC Convergence', in: D. Baird, A. Nordmann & J. Schummer (eds.), *Discovering the Nanoscale*, Amsterdam: IOS Press, pp. 21-33.

Khushf, G.: 2004, 'Systems Theory and the Ethics of Human Enhancement: A Framework for NBIC Convergence', *Annals of the New York Academy of Sciences*, 1013, 124-149.

Krätschmer, W., L.D. Lamb, K. Fostiropoulos, and D.R. Huffman. 1990. *Nature* 347:354.

Kresge, C.T., M.E. Leonowicz, W.J. Roth, J.C. Vartuli, and J.S. Beck. 1992. *Nature* 359:710.

Kupperman, A., S. Nadimi, S. Oliver, G. Ozin, J. Garcés, and M. Olken. 1993. *Nature* 365:239.

Kuusi, O.; Meyer, M.: 2002, 'Technological generalizations and leitbilder ' the anticipation of technological opportunities', *Technological Forecasting & Social Change*, 69, 625-639.

Landon, B.: 2004, 'Less is More: Much Less is Much More: The Insistent Allure of Nanotechnology Narratives in Science Fiction', in: N.K. Hayles (ed.), *Nanoculture: Implications of the New Technoscience*, Bristol, UK: Intellect Books, pp. 131-146.

Laszlo, P.: 2004, 'Is There Life After Partington?', *Hyle: International Journal for Philosophy of Chemistry*, 10(2), 169-178.

Lenhard, J.: 2004, 'Nanoscience and the Janus-Faced Character of Simulations', in: D. Baird, A. Nordmann & J. Schummer (eds.), *Discovering the Nanoscale*, Amsterdam: IOS Press, pp. 93-100.

Lent, C.S., Tougaw, P.D., Porod, W., Bernstein, G.H., "Quantum Cellular Automata," *Nanotechnology*, Vol. 4, p. 49, 1993.

Lewak, S.E.: 2004, 'What's the Buzz? Tell Me What's A-Happening: Wonder, Nanotechnology, and Alice's Adventures in Wonderland', in: N.K. Hayles (ed.), *Nanoculture: Implications of the New Technoscience*, Bristol, UK: Intellect Books, pp. 201-.

Lide, D.R., ed. 1993-1994. *CRC Handbook of Chemistry and Physics*, 74th ed.

Mehta, M.D.: 2002, 'Nanoscience and Nanotechnology: Assessing the Nature of Innovation in These Fields', *Bulletin of Science, Technology & Society*, 22(4), 269-273.

Mehta, M.D.: 2004, 'From Biotechnology to Nanotechnology: What Can We Learn from Earlier Technologies?, *Bulletin of Science, Technology & Society*, 24, 34-39.

Meyer, M. & Kuusi, O.: 2004, 'Nanotechnology: Generalizations in an Interdisciplinary Field of Science and Technology', *Hyle: International Journal for Philosophy of Chemistry*, 10(2), 153-168.

Meyer, M.: 2000a, 'Does science push technology? Patents citing scientific literature', *Research Policy*, 29, 409-434.

Meyer, M.: 2000b, 'Patent citations in a novel field of technology: What can they tell about interactions of emerging communities of science and technology?', *Scientometrics*, 48, 151-178.

Meyer, M.: 2001a, 'Patent citations in a novel field of technology: An exploration of nano-science and nano-technology', *Scientometrics*, 51, 163-183.

Meyer, M.: 2001b, 'Socio-economic Research on Nanoscale Science and Technology: A European Overview and Illustration', in: M.C. Roco and W.S. Bainbridge (eds.), *Societal Implications of Nanoscience and Nanotechnology*, Dordrecht: Kluwer, pp. 217-241.

Milburn, C.: 2002, 'Nanotechnology in the age of post-human engineering: science fiction as science', *Configurations*, 10, 261-295 [reprinted in: N.K. Hayles (ed.), *Nanoculture: Implications of the New Technoscience*, Bristol, UK: Intellect Books, 2004, pp. 109-130].

Milburn, C.: 2004, 'Nano/Splatter: Disintegrating the Postbiological Body', *New Literary History*, 35 (in print).

Mnyusiwalla, A.; Abdallah, S.D.; Singer, P.A.: 2003, 'Mind the gap: science and ethics in nanotechnology', *Nanotechnology*, 14, R9-R13.

Mody, C.C.M.: 2004, 'How Probe Microscopists Became Nanotechnologists', in: D. Baird, A. Nordmann & J. Schummer (eds.), *Discovering the Nanoscale*, Amsterdam: IOS Press, pp. 119-133.

Nordmann, A.: 2004, 'Nanotechnology: Convergence and Integration', Presentation at *EuroNanoForum, Trieste, December 10, 2003* (Proceedings in preparation).

Nordmann, A.: 2004, 'Nanotechnology's WorldView: New Space for Old Cosmologies', *IEEE Technology and Society Magazine*, 23 (forthcoming).

Nordmann, A.: 2004, 'Social Imagination for Nanotechnology', in: European Commission (Community Health and Consumer Protection): *Nanotechnologies: A Preliminary Risk Analysis on the Basis of a Workshop, Brussels, 1-2 March 2004*, pp. 111-113 [www.europa.eu.int/comm/health/ph_risk/documents/ev_20040301_en.pdf].

Nordmann, A.: 2004, 'Was ist TechnoWissenschaft? ' Zum Wandel der Wissenschaftskultur am Beispiel von Nanoforschung und Bionik', in: T. Rossmann & C. Tropea (eds.), *Bionik ' Neue Forschungsergebnisse aus Natur-, Ingenieur- und Geisteswissenschaften*, Berlin: Springer, 2004 (forthcoming)

Paschen, H.; Coenen, C.; Fleischer, T.; Grünwald, R.; Oertel, D. & Revermann, C.: 2004, *Nanotechnologie: Forschung, Entwicklung, Anwendung*, Berlin: Springer (*Nanotechnologie, TAB-Arbeitsbericht 92*, Berlin: Büro für Technikfolgen-Abschätzung beim Deutschen Bundestag).

Pitt, J.C.: 2004, 'The Epistemology of the Very Small', in: D. Baird, A. Nordmann & J. Schummer (eds.), *Discovering the Nanoscale*, Amsterdam: IOS Press, pp. 157-163.

Pressman, J.: 2004, 'Nano Narrative: A Parable from Electronic Literature', in: N.K. Hayles (ed.), *Nanoculture: Implications of the New Technoscience*, Bristol, UK: Intellect Books, pp. 191-200.

Prigogine, I., and S. Rice. 1988. *Advances in chemical physics*, Vol. 70, Parts 1 & 2. New York: J. Wiley.

R. T. Bate, "Nanoelectronics," *Nanotechnology*, Vol. 1, pp. 1-7, 1990.

Rademann, K., B. Kaiser, U. Even, F. Hensel. 1987. *Phys. Rev. Lett.* 59:2319.

Rao, C.N.R., B.C. Satishkumar, and A. Govindaraj. 1997. *Chem. Commun.* 1581.

Rao, M.B., and S. Sircar. 1993. *Gas Separation and Purification* 7:279.

Reed, M.A., "Quantum Dots," *Scientific American,* January 1993, pp. 118-123.

Roberts, J.A.: 2004, 'Deciding the Future of Nanotechnologies: Legal Perspectives on Issues of Democracy and Technology', in: D. Baird, A. Nordmann & J. Schummer (eds.), *Discovering the Nanoscale*, Amsterdam: IOS Press, pp. 247-255.

Robinson, C.: 2004, 'Images in NanoScience/Technology', in: D. Baird, A. Nordmann & J. Schummer (eds.), *Discovering the Nanoscale*, Amsterdam: IOS Press, pp. 165-169.

Robison, W.L.: 2004, 'Nano-Ethics', in: D. Baird, A. Nordmann & J. Schummer (eds.), *Discovering the Nanoscale*, Amsterdam: IOS Press, pp. 285-300.

Roco, M.C. & Bainbridge, W.S. (eds.): 2001, *Societal implications of nanoscience and nanotechnology*, (Proceedings of a workshop organized by the National Science Foundation, September 28-29, 2000), Kluwer: Dordrecht [available online at http://itri.loyola.edu/nano/societalimpact/nanosi.pdf]

Ruthven, D.M., S. Farooq, K.S. Knaebel. 1994. *Pressure swing adsorption*. New York: VCH Publishers.

Sarewitz, D.; Woodhouse, E.: 2003, 'Small is Powerful', in: A. Lightman, D. Sarewitz & Chr. Desser, (eds.), *Living with the Genie: Essays on Technology and the Quest for Human Mastery*, Washington, DC: Island Press, pp. 63-83.

Schiemann, G.: 2004, 'Dissolution of the Nature-Technology Dichotomy? Perspectives on Nanotechnology from an Everyday Understanding of Nature', in: D. Baird, A. Nordmann & J. Schummer (eds.), *Discovering the Nanoscale*, Amsterdam: IOS Press, pp. 209-213.

Schmidt, J.C.: 2004, 'Unbounded Technologies: Working Through Technological Reductionism of Nanotechnology', in: D. Baird, A. Nordmann & J. Schummer (eds.), *Discovering the Nanoscale*, Amsterdam: IOS Press, pp. 35-50.

Schummer, J.: 2004, 'Why do Chemists Perform Experiments?', in: D. Sobczynska, P. Zeidler, E. Zielonacka-Lis (eds.), *Chemistry in the Philosophical Melting Pot*, Peter Lang (Frankfurt/M.), 2004, pp. 395-410.

Schummer, J.: 2005, 'Reading Nano: The Public Interest in Nanotechnology as Reflected in Book Purchase Patterns", *Public Understanding of Science*, 14 (2) (forthcoming).

Schummer, J.; Baird, D. (eds.): 2004-5, *Nanotech Challenges*, joint special issue of *Hyle: International Journal for Philosophy of Chemistry & Techne: Research in Philosophy and Technology* (forthcoming).

Stuckless, J.T, D.E. Starr, D.J. Bald, C.T. Campbell. 1997. *J. Chem. Phys.* 107:5547.

Suchman, M.: 2001, 'Envisioning Life on the Nano-Frontier', in: M.C. Roco and W.S. Bainbridge (eds.), *Societal Implications of Nanoscience and Nanotechnology*, Dordrecht: Kluwer, pp. 211-216.

Index